Copyright©S. Nakamura 2016
All Rights Reserved
ISBN-13: 978-1530228027
ISBN-10: 1530228026
Date of Publication: February 2016
LCCN: 2016903653
Publisher: CreateSpace
North Charleston, SC

GNU Octave Matlab Tutorial Series Vol. 2

Foundation of Numerical Analysis
Implementation with GNU Octave/Matlab
Edition 2.0

S. Nakamura

Printed by
CreateSpace
7290 Investment Dr. Suite B
North Charleston, SC 29418, USA

—

About the picture on the front cover

The picture on the front cover was created at NASA Goddard Space Flight Center Visualization Studio. It shows Atlantic Ocean flow viewed from a satellite. On the left side of the picture the eastern seaboard of North America including Saint Pierre, Nova Scotia, Rhode Island, Long Island, and Chesapeake Bay is seen. It is an amazing picture of the vortices caused by the jet-like ocean stream coming from Gulf of Mexico turning around the south tip of Florida compounded with the effect of the Earth's rotation. Although no detail is available on how this visualization was created, author believes that it is not a direct camera shot or measurement of stream lines from a satellite but rather that ocean flow velocity distribution was first measured from the satellite, and then integrated by a *numerical integration method* to create a stream-line map.

Table of Contents

Preface

Numerical methods are used in almost every technology today. If you are a college student in science or engineering discipline, to learn methods of numerical analysis at an early stage of your student carrier will be highly beneficial. This is because numerical methods cover most of mathematics that students learn in high school and college math courses. In learning numerical methods, students will practice mathematics significantly more thoroughly than in math courses. It will also enhance your ability to understand higher level mathematics which you will likely be exposed.

This book is intended to be a college level introductory numerical analysis course text as well as a reference book even when the student uses other books as a primary text. The objective is for the readers to understand comfortably and quickly the fundamentals of nume-rical methods and become able to solve mathematical problems by numerical methods using GNU Octave or Matlab. It is assumed that the reader has either Matlab or GNU Octave as mathematical software for numerical computations. The students in US colleges may have a student edition of Matlab. If neither is available at this time, the reader is suggested to obtain GNU Octave as it is equivalent to Matlab, and can be down-loaded for free of any charge from the following site:
https://www.gnu.org/software/octave/download.html

In writing this book, author spent much effort to make the book easy to understand and least boring. Each subject is started with simple examples and then explained in more general terms, all with illustrative applications of Octave/Matlab and graphic expl-anations.

To use GNU Octave or Matlab proficiently the reader should have basic knowledge of commands and graphical tools of Octave/ Matlab. If not, please read the author's recently published book, *Octave/Matlab Primer and Applications*, or *GNU Octave Primer for beginners*, both of which are seamlessly connected to this book.

—

Print and Kindle versions of the latter are available at a low cost at Amazon.com. These books are good as a tutorial for Matlab users too because of equivalence between the two software.

Reading mathematical texts is often boring and readers might get sleepy. The author hopes this book is not that kind because reading this book requires reader's work on Octave or Matlab. The numerical methods you read should be applied with Octave/Matlab, which requires motion of your fingers, which will brown away sleepiness. The readers might also create their own exercise problems and solve them on Octave/ Matlab.

Readers are encouraged to visit **http://octave.ismr.us** that is a companion website for this book. Octave/Matlab scripts in this book are all listed in this website, which can be copy/pasted to Octave/Matlab editor without retyping. Additional information including corrections if any will be posted on the site.

Chapter 1
Linear Algebra

This chapter describes basic aspects of solving linear equations. The reader will learn simple but powerful means of solving linear equations available in Octave/Matlab software. We also learn that certain linear equations cannot be solved, or if solvable, the solutions can be very inaccurate. We regard the knowledge of these aspects as important because we need to be cautious to the results of the calculations that can be done so easily with few simple commands. Toward the end of this chapter, we will explain the limitations of Octave and Matlab, but also give brief directions on what to do when the number of unknowns is huge.

Octave/Matlab commands used in this chapter:
eye(n): n-by-n identity matrix where n is an integer
inv(a): inversion of matrix a
cond(a): condition number of matrix a
det(a): determinant of matrix a
eig(a): eigenvalues of matrix a
hilb(n): n-by-n Hilbert matrix
poly(a): characteristic polynomial of matrix a
poly(r): normalized polynomial with roots r in array
roots(c): roots of polynomial, where c is an array of coefficients
roots(poly(a)): eigenvalues of matrix a
$a*b$: product of matrices a and b,
   or product of matrix a and vector b
[$c_1, c_2, c_3, ...$]: definition of row vector
[$c_1; c_2; c_3; ...$] : definition of column vector
[$c_1, c_2, c_3, ...$]': definition of column vector
[$c_{1,1}, c_{1,2}, c_{1,3}; c_{2,1}, c_{2,2}, c_{2,3}; ...$]: definition of a matrix
**** : inverse division operator

1.1 Arithmetic rules of array variables

Before we study linear algebra in this chapter, it is imperative to understand how multiplication, addition and subtraction of array variables work.

We start with a 2x2 two-dimensional array variable:

$$c = \begin{pmatrix} c_{1,1} & c_{1,2} \\ c_{2,1} & c_{2,2} \end{pmatrix}$$

(1.1)

a vertical one-dimensional array:

$$s = \begin{pmatrix} s_1 \\ s_2 \end{pmatrix}$$

(1.2)

and a row one-dimensional array:

$$s = \begin{pmatrix} s_1 & s_2 \end{pmatrix}$$

(1.3)

In this chapter we call the two-dimensional array variable like c as matrix, and the one-dimensional array like s as vector. Matrix and vector are standard names of arrays in linear algebra. In general, matrices do not have to be square arrays, although matrices involved in calculations in this chapter are all square. Vectors may be considered as special forms of matrix. Indeed, a column vector is a matrix with only one column, and a row vector is a matrix with only one row.

The product of a matrix c and a column vector s is written as cs in mathematical notations and yields a column vector:

$$cs = \begin{pmatrix} c_{1,1}s_1 + c_{1,2}s_2 \\ c_{2,1}s_1 + c_{2,2}s_2 \end{pmatrix}$$

(1.4)

For example, if

$$c = \begin{pmatrix} 1 & 3 \\ 2 & -1 \end{pmatrix}$$

(1.5)

and

$$s = \begin{pmatrix} 2 \\ 1 \end{pmatrix} \tag{1.6}$$

then

$$cs = \begin{pmatrix} 1*2+3*1 \\ 2*2-1*1 \end{pmatrix} = \begin{pmatrix} 5 \\ 3 \end{pmatrix} \tag{1.7}$$

In Octave/Matlab, the foregoing calculations are done as follows:

>>c = [1, 3; 2, -1]; s = [2; 1]; d = a*s

then, we get the following result

d =
  5
  3

Let us consider another matrix:

$$e = \begin{pmatrix} e_{1,1} & e_{1,2} \\ e_{2,1} & e_{2,2} \end{pmatrix} \tag{1.8}$$

The product of c and e is ce, which is a matrix:

$$ce = \begin{pmatrix} c_{1,1}e_{1,1} + c_{1,2}e_{2,1}, & c_{1,1}e_{1,2} + a_{1,2}e_{2,2} \\ c_{2,1}e_{1,1} + c_{2,2}e_{2,1}, & c_{2,1}e_{1,2} + c_{2,2}e_{2,2} \end{pmatrix} \tag{1.9}$$

If e is given in numbers by

$$e = \begin{pmatrix} 7 & -1 \\ -2 & 3 \end{pmatrix} \tag{1.10}$$

then, the Octave/Matlab calculation of the product is

>>c = [1, 3; 2, -1]; e = [7, -1; -2, 3]; g = c*e

g =
```
 1   8
16  -5
```

The readers should verify this by hand calculations, or using Octave/Matlab.

The same rule extends to any larger matrices and vectors. A general expression of multiplication of an n-by-n matrix and an n column vector is

$$cs = \begin{pmatrix} \sum_k c_{1,k} s_k \\ \sum_k c_{2,k} s_k \\ .. \\ \sum_k c_{n,k} s_k \end{pmatrix} \qquad (1.11)$$

where

$$c = \begin{pmatrix} c_{1,1} & c_{1,2} & .. & c_{1,n} \\ c_{2,1} & c_{2,2} & .. & c_{2,n} \\ .. & .. & .. & .. \\ c_{n,1} & c_{n,2} & .. & c_{n,n} \end{pmatrix} \qquad (1.12)$$

$$s = \begin{pmatrix} s_1 \\ s_2 \\ .. \\ s_n \end{pmatrix} \quad \text{or equivalently} \quad s = \begin{pmatrix} s_1 & s_2 & .. & s_n \end{pmatrix}' \qquad (1.13)$$

and where the sign ' is an operator to transpose a row vector to a column vector, or vise versa. Equation (1.11) may be written more compactly as

$$cs = \left(\sum_k c_{i,k} s_k \right) \tag{1.14}$$

with

$$c = \left(c_{i,j} \right)$$

$$\tag{1.15}$$

$$s = \left(s_{i,1} \right)$$

Note here that, when two indices are attached to each element in a matrix, the first index changes vertically downward, and the second index changes horizontally to the right.

The product of an *n*-by-*n* matrix *a* and another *n*-by-*n* matrix *b* is

$$ab = \begin{pmatrix} \sum_k a_{1,k} b_{k,1} & \sum_k a_{1,k} b_{k,2} & .. & \sum_k a_{1,k} b_{k,n} \\ \sum_k a_{2,k} b_{k,1} & \sum_k a_{2,k} b_{k,2} & .. & \sum_k a_{2,k} b_{k,n} \\ .. & .. & .. & .. \\ \sum_k a_{n,k} b_{k,1} & \sum_k a_{n,k} b_{k,2} & & \sum_k a_{n,k} b_{k,n} \end{pmatrix} \tag{1.16}$$

or equivalently and more compactly

$$ab = \left(\sum_k a_{i,k} b_{k,j} \right)$$

where

$$a = \begin{pmatrix} a_{1,1} & a_{1,2} & .. & a_{1,n} \\ a_{2,1} & a_{2,2} & .. & a_{2,n} \\ .. & .. & .. & .. \\ a_{n,1} & a_{n,2} & & a_{n,n} \end{pmatrix} \tag{1.17}$$

$$b = \begin{pmatrix} b_{1,1} & b_{1,2} & .. & b_{1,n} \\ b_{2,1} & b_{2,2} & .. & b_{2,n} \\ .. & .. & .. & .. \\ b_{n,1} & b_{n,2} & & b_{n,n} \end{pmatrix}$$
(1.18)

Addition or subtraction of two vectors or matrices works when the two arrays are in the same size and same shape: both are column vectors of the same length, or both are row vectors of the same length, or both are matrices of the same size. Operation of addition or subtraction affects each element in the same position in the array variables. For example, if

$$a = \begin{pmatrix} 1 & 3 \\ 2 & -1 \end{pmatrix}, \quad b = \begin{pmatrix} 7 & 3 \\ 4 & 1 \end{pmatrix}$$
(1.19)

then

$$a \pm b = \begin{pmatrix} 1 \pm 7 & 3 \pm 3 \\ 2 \pm 4 & -1 \pm 1 \end{pmatrix}$$
(1.20)

If

$$a = \begin{pmatrix} 2 \\ 1 \end{pmatrix}, \quad b = \begin{pmatrix} 1 \\ 8 \end{pmatrix}$$
(1.21)

then

$$a + b = \begin{pmatrix} 2+1 \\ 1+8 \end{pmatrix} = \begin{pmatrix} 3 \\ 9 \end{pmatrix}$$
(1.22)

Also, if

$$a = \begin{pmatrix} 2 & 1 \end{pmatrix}, \quad b = \begin{pmatrix} 4 & 5 \end{pmatrix}$$
(1.23)

then

$$a+b=(2+4 \quad 1+5)=(6 \quad 6) \tag{1.24}$$

In more general terms, the addition and subtraction rules may be written as

$$a \pm b = \left(a_{i,j}\right) \pm \left(b_{i,j}\right) = \left(a_{i,j} \pm b_{i,j}\right) \tag{1.25}$$
$$i = 1,2,...,m \ \text{and} \ j = 1,2,...,n$$

where m and n are respectively height and width of the array variables, a and b.

Example
We practice the multiplication and addition/subtraction rules with Octave/Matlab using 3x3 matrices and 3 vectors. We first define two matrices and two vectors in mathematical term:

$$c = \begin{pmatrix} 1 & 0 & 2 \\ 2 & 1 & 1 \\ 1 & 3 & 0 \end{pmatrix} \tag{a}$$

$$e = \begin{pmatrix} 0 & 2 & 2 \\ 4 & 1 & 0 \\ 1 & 0 & 5 \end{pmatrix} \tag{b}$$

$$s = \begin{pmatrix} 1 \\ 1 \\ 2 \end{pmatrix} \tag{c}$$

$$t = \begin{pmatrix} 4 \\ 5 \\ 1 \end{pmatrix} \tag{d}$$

The calculation of $g = cs$ in Octave/Matlab is

```
>>c = [1 0 2; 2 1 1; 1 3 0]; s = [1;1;2];  g = c*s
g =
   5
   5
   4
```

The calculation of $h = ce$ is

```
>>c = [1 0 2; 2 1 1; 1 3 0]; e = [0 2 2; 4 1 0; 1 0 5]; h = c*e
h =
    2   2  12
    5   5   9
   12   5   2
```

The calculation of $c + e$ is

```
>>c = [1 0 2; 2 1 1; 1 3 0]; e = [0 2 2; 4 1 0; 1 0 5]; h = c+e
h =
   1  2  4
   6  2  1
   2  3  5
```

The calculation of $s + t$ is

```
>>s=[1 1 2]'; t=[4 5 1]'; u=s+t
u=
   5
   6
   3
```

The above command line may be equivalently written as

```
>>s=[1; 1; 2]; t=[4; 5; 1]; u=s+t
```

Readers are encouraged to verify at least a few numbers in h by hand calculations.

1.2 Special matrices

Diagonal matrix
A diagonal matrix is a matrix such that all elements are zero except the diagonal elements, for example

$$D = \begin{pmatrix} 3 & 0 & 0 \\ 0 & 1 & 0 \\ 0 & 0 & 2 \end{pmatrix} \tag{1.26}$$

Identity matrix
An identity matrix, denoted by I, is a diagonal matrix in which its diagonal elements are all unity but other elements are all zero. The following matrices are all identity matrices:

$$I = \begin{pmatrix} 1 & 0 \\ 0 & 1 \end{pmatrix}, \quad I = \begin{pmatrix} 1 & 0 & 0 \\ 0 & 1 & 0 \\ 0 & 0 & 1 \end{pmatrix}, \quad I = \begin{pmatrix} 1 & 0 & 0 & 0 \\ 0 & 1 & 0 & 0 \\ 0 & 0 & 1 & 0 \\ 0 & 0 & 0 & 1 \end{pmatrix}, \quad \dots \tag{1.27}$$

The identity matrices have the following property. Assuming a is a matrix with the same size as I, the products of a and I become

$$aI = a, \quad Ia = a \tag{1.28}$$

Neither post-multiplication or pre-multiplication changes the matrix a. Also, for a column vector s, pre-multiplication of I does not alter s, namely $Is = s$.

In Octave/Matlab, an identiy matrix is **eye(n)** where n is the size of the matrix, for example

```
>> eye(3)
ans =
   1  0  0
   0  1  0
```

0 0 1

Inverse matrix

If the product of two matrices, a and b of the same size satisfy

$$ab = I \qquad\qquad (1.29)$$

then a is called inverse of b, and b is inverse of a. If a and b are inverse to each other, the following equation is also true

$$ba = I \qquad\qquad (1.30)$$

Inverse of a is expressed also as a^{-1}, and inverse of b as b^{-1}. Therefore the following equations are all true:

$$aa^{-1} = a^{-1}a = bb^{-1} = b^{-1}b = I \qquad\qquad (1.31)$$

Inverse of a matrix may be obtained on Octave/Matlab by the **inv** command. For example, inverse of

$$a = \begin{pmatrix} 1 & 4 & 0 \\ 2 & 2 & 1 \\ 3 & 0 & 1 \end{pmatrix} \qquad\qquad (1.32)$$

is calculated by

>> a = [1 4 0; 2 2 1; 3 0 1]; b = inv(a)

b =
  0.33333 -0.66667 0.66667
  0.16667 0.16667 -0.16667
 -1.00000 2.00000 -1.00000

This can be verified by

>> a*b

ans =

$$\begin{matrix} 1 & 0 & 0 \\ 0 & 1 & 0 \\ 0 & 0 & 1 \end{matrix}$$

Singular matrix

A matrix is singular if at least one of the rows can be eliminated by adding or subtracting multiple of other rows. For example, the matrix given by

$$b = \begin{pmatrix} 1 & 4 & 1 \\ 2 & 2 & 1 \\ 3 & 0 & 1 \end{pmatrix} \tag{1.33}$$

is singular. In fact, in matrix b, subtract 2 times the 2nd row from the 1st row and add the 3rd row to the 1st row, then the 1st row becomes all zeros.

If inversion of a singular matrix is attempted, Octave/Matlab response is:

```
>> b = [ 1 4 1; 2 2 1; 3 0 1];  c = inv(b)
```

warning: matrix singular to machine precision:
c =
  Inf Inf Inf
  Inf Inf Inf
  Inf Inf Inf

The printout says that matrix b is singular so no inverse matrix exists.

Symmetric matrix

We define matrix a by

$$a = \left(a_{i,j} \right) \tag{1.34}$$

If

$$a_{i,j} = a_{j,i} \tag{1.35}$$

then matrix a is said to be symmetric. The following examples are all symmetric matrices:

$$\begin{pmatrix} 1 & 7 \\ 7 & 2 \end{pmatrix}, \begin{pmatrix} 2 & 0 & 4 \\ 0 & 1 & 3 \\ 4 & 3 & 6 \end{pmatrix}, \begin{pmatrix} 4 & 5 & 1 & 9 \\ 5 & 2 & 0 & 2 \\ 1 & 0 & 3 & 7 \\ 9 & 2 & 7 & 1 \end{pmatrix}, \begin{pmatrix} a_{1,1} & a_{2,1} & .. & a_{n,1} \\ a_{2,1} & a_{2,2} & .. & a_{n,2} \\ .. & .. & .. & .. \\ a_{n,1} & a_{n,2} & .. & a_{n,n} \end{pmatrix} \quad (1.36)$$

Diagonal matrices and identity matrices are symmetric matrices. Notice also, if $a=a'$, the matrix a is symmetric.

1.3 Linear equations

The simplest linear equation has one unknown variable x in the equation as

$$ax = s \qquad (1.37)$$

where a and s are constants. Its solution is immediately $x = s/a$, or equivalently $x = a^{-1}s$, although the author may be blamed for a bit unusual expression of the latter.

Next is a set of two equations that has two variables, x_1 and x_2:

$$\begin{aligned} a_{1,1}x_1 + a_{1,2}x_2 &= s_1 \\ a_{2,1}x_1 + a_{2,2}x_2 &= s_2 \end{aligned} \qquad (1.38)$$

where a's and s's are constants. A graphic interpretation of the foregoing equation is as follows. On a two-dimensional coordinate, each equation represents a line as illustrated in Figure 1.1. The intersection of the two lines is the solution.

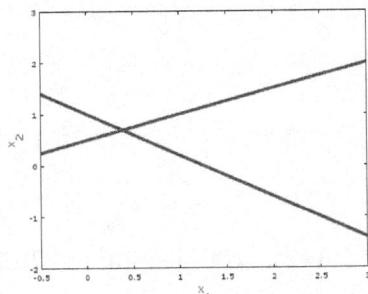

Figure 1.1   Plot of two linear equations

Equation (1.38) is equivalently written as

$$ax = s \qquad (1.39)$$

where a is a 2x2 matrix,

$$a = \begin{pmatrix} a_{1,1} & a_{1,2} \\ a_{2,1} & a_{2,2} \end{pmatrix} \qquad (1.40)$$

and x and s are column vectors:

$$x = \begin{pmatrix} x_1 \\ x_2 \end{pmatrix}, \quad s = \begin{pmatrix} s_1 \\ s_2 \end{pmatrix}$$

Pre-multiplying Eq.(1.39) by a^{-1} yields

$$x = a^{-1}s$$

that is, the solution of the linear equation equals pre-multiplication of s by a^{-1}.

To illustrate solution of a linear equation with Octave/Matlab, we assume

$$a = \begin{pmatrix} 3 & 1 \\ 2 & 5 \end{pmatrix} \qquad (1.41)$$

and

$$s = \begin{pmatrix} 2 \\ 1 \end{pmatrix}$$
(1.42)

The inverse of a is obtained by **inv(a)**, so the solution of the linear equation becomes

```
>> a = [3 1; 2 5]; s=[2 1]'; x=inv(a)*s
x =
  0.692308
 -0.076923
```

However, an equivalent but better way is

```
>> a=[3 1; 2 5], s=[2 1]', x=a\s.

x =
  0.692308
 -0.076923
```

The operator \ in the foregoing script is named "inverse division operator." The results of **x=inv(a)*s** and **x=a\s** are identical but efficiency of computation of the latter is better. The difference becomes clear in terms of computational speed when the matrix size becomes large.

Although we have explained solution of linear equation using a very small number of unknowns, the principle is the same regardless to the size of the problems. For one more numerical illustration, we assume

$$a = \begin{pmatrix} 3 & 2 & 0 \\ 1 & 2 & 5 \\ 9 & 1 & 2 \end{pmatrix}$$
(1.43)

$$s = \begin{pmatrix} 2 \\ 3 \\ 1 \end{pmatrix} \qquad\qquad (1.44)$$

The Octave/Matlab calculation of the solution is

```
>> a = [3 2 0; 1 2 5; 9 1 2]; s = [2; 3; 1]; x=a\s
x =
 -0.048193
  1.072289
  0.180723
```

The foregoing command line may be equivalently written as

```
>> x = [3 2 0; 1 2 5; 9 1 2]\ [2; 3; 1]
```

or

```
>> x = [3 2 0; 1 2 5; 9 1 2]\ [2 3 1]'
```

Both of which yield the same result.

Although we do not explain in this chapter how Octave/ Matlab solves linear equations, we note that the Gauss elimination method is indeed used. Details of the Gauss elimination method are left to other text books such as [Nakamura, 2002].

1.4 Determinant

A quantity named *determinant* is associated with any given matrix, a, and expressed by $\det(a)$. For a 2x2 matrix

$$a = \begin{pmatrix} a_{1,1} & a_{1,2} \\ a_{2,1} & a_{2,2} \end{pmatrix} \qquad\qquad (1.45)$$

the determinant can be written explicitly as

$$\det(a) = a_{1,1}a_{2,2} - a_{1,2}a_{2,1} \qquad (1.46)$$

For a 3x3 matrix

$$a = \begin{pmatrix} a_{1,1} & a_{1,2} & a_{1,3} \\ a_{2,1} & a_{2,2} & a_{2,3} \\ a_{3,1} & a_{3,2} & a_{3,3} \end{pmatrix} \qquad (1.47)$$

the determinant can be expanded as

$$\det(a) = a_{1,1}a_{2,2}a_{3,3} + a_{1,2}a_{2,3}a_{3,1} + a_{1,3}a_{2,1}a_{3,2}$$
$$-a_{1,3}a_{2,2}a_{3,1} - a_{1,2}a_{2,1}a_{3,3} - a_{1,1}a_{2,3}a_{3,2} \qquad (1.48)$$

For larger matrices, the determinant can be expanded and written down explicitly, but it will be very long and cumbersome. Rather we should know how to compute the determinant of a given matrix in Octave/Matlab. For any given matrix a, it is simply

>> det(a)

For an example of

$$a = \begin{pmatrix} 3 & 2 & 0 \\ 1 & 2 & 5 \\ 9 & 1 & 2 \end{pmatrix} \qquad (1.49)$$

calculation of its determinant is

>> a = [3 2 0; 1 2 5; 9 1 2]; det(a)
ans = 83

What is the meaning of determinant is a good question. The particular value of the determinant 83 in the foregoing example has no specific meaning. The value of determinant becomes important only when it becomes zero, or extremely large or extremely small.

1.4 Singular problems

Not every linear equation has a solution. We explain why by a geometrical interpretation first. We earlier explained that two equations in a 2x2 problem are 2 lines if plotted on the 2-dimensional coordinates as illustrated in Fig 1.1.

Now, consider the following 2 linear equations,

$$2x_1 + x_2 = 1$$
$$4x_1 + 2x_2 = 3$$

(1.50)

The two equations may look different, but a closer look at reveals that the left side of one equation equals a constant times the left side of the other equation. It means that when plotted, the two lines become parallel to each other as illustrated in Figure 1.2, so there is no intersection or solution. When the foregoing equation is written in a matrix form, the matrix of the coefficients becomes singular and its determinant becomes zero.

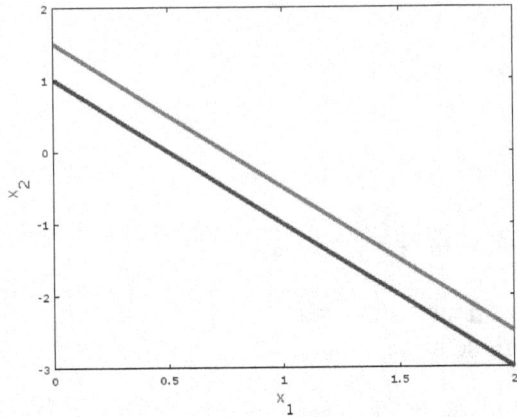

Figure 1.2 Graphic illustration of a set
of singular linear equations

For any larger number of unknowns, the linear equation can become singular too. The problem is singular if the left side of one equation equals a multiple of another equation, or summation of multiple of other equations. Such singular problems have two symptoms. First, if solution of such a problem is attempted, Octave/Matlab gives a warning that ***the problem is singular***. Second is that the determinant computed by the **det** command becomes zero, for example

>> det([1 1 1; 1 2 3; 2 3 4])
ans = 0

where the 3rd row equals 1st row plus 2nd row.

1.5 Pathological problems

Pathological problem refers to a set of linear equations that has a solution, yet the solution becomes inaccurate, or the solution is significantly affected when some of the elements of the matrix or the source term change slightly. Why this pathological case occurs can be explained graphically if we consider a 2x2 problem.

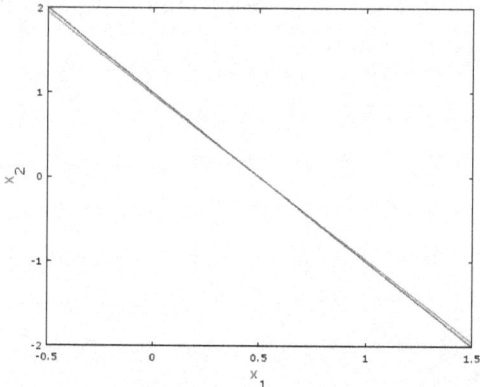

Figure 1.3   Plot of two linear equations
with very close slopes

In Figure 1.3, two lines are plotted, which are close to each other with a very small difference in slopes. There is an intersection, which is the solution, but a small change in a coefficient or the right side terms may shift the location of the intersection significantly. It also means, even if the constants are not intentionally changed, round off errors during the computation may

---

24

cause a similar effect. This is the pathological symptom of a linear equation.

Graphical explanation of the pathological behavior of a larger set of linear equations is not easy, but the nature of problem is similar, except it happens in more complex and severe ways. The solutions of pathological problems are inaccurate and significantly affected by precision of the computer software.

The matrix that causes a pathological behavior is called an ill-conditioned matrix. The determinant of an ill-conditioned matrix tends to be very small. Another and better indicator for ill-conditioned matrices is the condition number defined by

$$\text{cond}(a) = \|a\| \|a^{-1}\| \tag{1.51}$$

where we use the definition given by

$$\|a\| = \sqrt{\lambda_{max}(a'a)} \tag{1.52}$$

named the 2-norm, or spectral norm, and where $\lambda_{max}(a'a)$ is the largest eigenvalue of $a'a$, where a' is transpose of a. In Octave/ Matlab the condition number is computed by

>> cond(a)

The condition number quantifies the sensitivity of the inverse matrix to the round-off errors during the calculation of the inverse matrix and tends to be huge for ill-conditioned matrices. See Appendix E for more explanations of norm and condition number in Octave/Matlab.

The following set of 2 linear equations is an example of ill-conditioned problem:

$$0.12065x_1 + 0.98775x_2 = 2.01045$$
$$0.12032x_1 + 0.98755x_2 = 2.00555 \tag{1.53}$$

The condition number for this problem is calculated by
>>a=[0.12065 0.98775; 0.12032 0.98755]; cond(a)
ans=6559.8

Solution of Eq.(1.53) with Octve/Matlab is:
>> a=[0.12065 0.98775; 0.12032 0.98755]\[2.01045, 2.00555]'
a =
  14.703381898600997
   0.239419867308317

However, if Fortran is used in single precision, the solution is

  14.70332
   0.2394276

Here the difference of Fortran results appears in the 5th decimal place. For this problem, Octave/Matlab result is more credible because of its double precision and the size of the matrix is small. Although the difference between these two solutions is relatively small, if a problem that has a greater condition number is solved, the difference would be significantly larger.

Hilbert matrices defined by

$$a_{i,j} = \frac{1}{i+j-1}$$
(1.54)

are notoriously ill-conditioned. In the following table, n is size of Hilbert matrices, $\det(a)$ is the determinant, and $\text{cond}(a)$ is the condition numbers for $n=2$ to 14. As n increases, the determinant decreases rapidly, and the condition number increases to a huge value.

n	$\det(a)$	$\text{cond}(a)$
2	0.083333	1.9281e+001
3	4.6296e-004	5.2406e+002
4	1.6534e-007	1.5514e+004
5	3.7493e-012	4.7661e+005
6	5.3673e-018	1.4951e+007
7	4.8358e-025	4.7537e+008

```
8   2.7370e-033  1.5258e+010
9   9.7202e-043  4.9315e+011
10 2.1640e-053  1.6025e+013
11 3.0169e-065  5.2232e+014
12 2.6973e-078  1.7408e+016
13 4.2110e-094  1.9697e+018
14 -3.922e-107  5.1527e+017

% Hilbert matrix det(a) and cond(a)
clear all
for n=5:14
for i=1:n
for j=1:n
a(i,j)=1/(i+j-1);
end
end
[n, det(a), cond(a)]
end
```

Note that Hilbert matrix can be generated in Octave/Matlab simply by **hilb(n)** where n is the size of the matrix. This command is used in the next script.

To show how bad the solution of a linear equation could be when the coefficient matrix is very ill-conditioned, we test by solving

$$ax = s$$

where a is a Hilbert matrix, and s is a known vector. We will print out both x and ax after solution of the equation, so comparison of ax to s will show how agreement is good or bad. In the script and pint-out, ax is denoted by d.

In the first test, we assume that a is a 8-by-8 Hilbert matrix and

$$s = [0\ 0\ 0\ 0\ 0\ 0\ 0\ 1]'$$

Octave/Matlab script and its answers (Octave) are:

```
%C1_1.m
```

```
clear
n=8;
a=hilb(n);
[n, det(a),det(inv(a)), cond(a), cond(inv(a))]
s=[ 0 0 0 0 0 0 0 1 ]';
x=a\s;
d=a*x;
for i=1:n
fprintf('x(%i) = %e\n', i, x(i))
end %for
for i=1:n
fprintf('d(%i) = %e,  Exact= %f\n', i, d(i), s(i))
end %for

x(1) = -5.148001e+004
x(2) =  2.882881e+006
x(3) = -3.891889e+007
x(4) =  2.162160e+008
x(5) = -5.945941e+008
x(6) =  8.562155e+008
x(7) = -6.183779e+008
x(8) =  1.766794e+008

d(1) =  0.000000e+000,  Exact= 0.000000
d(2) = -3.725290e-009,  Exact= 0.000000
d(3) = -1.490116e-008,  Exact= 0.000000
d(4) =  1.862645e-009,  Exact= 0.000000
d(5) = -3.725290e-009,  Exact= 0.000000
d(6) =  1.303852e-008,  Exact= 0.000000
d(7) = -2.793968e-008,  Exact= 0.000000
d(8) =  1.000000e+000,  Exact= 1.000000
```

The accuracy of the result for the first test is not so bad, because the d values agree with the exact in seven decimal places. However, if single precision is used with Fortran or C, the results could be very poor.

In the second test, we assume that a is a 12-by-12 Hilbert matrix and

$$s = [0\ 0\ 0\ 0\ 0\ 0\ 0\ 0\ 0\ 0\ 0\ 1]'$$

The result (Octave) is as follows:

```
x(1) =   2.435382e+006
x(2) =  -2.496908e+008
x(3) =   6.239176e+009
x(4) =  -6.575896e+010
x(5) =   3.576606e+011
x(6) =  -1.086481e+012
x(7) =   1.821031e+012
x(8) =  -1.336652e+012
x(9) =  -5.055471e+011
x(10) = 1.746084e+012
x(11) = -1.246702e+012
x(12) = 3.103750e+011
```

```
d(1) =   0.0000190735,  Exact= 0.000000
d(2) =   0.0000114441,  Exact= 0.000000
d(3) =   0.0000343323,  Exact= 0.000000
d(4) =  -0.0000801086,  Exact= 0.000000
d(5) =   0.0007743835,  Exact= 0.000000
d(6) =  -0.0035552979,  Exact= 0.000000
d(7) =   0.0099182129,  Exact= 0.000000
d(8) =  -0.0180015564,  Exact= 0.000000
d(9) =   0.0211563110,  Exact= 0.000000
d(10) = -0.0155544281,  Exact= 0.000000
d(11) = 0.0065212250,  Exact= 0.000000
d(12) = 0.9988307953,  Exact= 1.000000
```

The result of the second test is poor, despite the use of double precision in Octave/Matlab, which is due to the ill-conditioned matrix. As the size of the Hilbert matrix increases, the result will become increasingly worse.

1.7 Eigenvalues of matrix

Eigenvalues of matrix a are solutions of the equation,

$$\det(a - \lambda I) = 0 \tag{1.55}$$

where λ, pronounced "lambda", is an eigenvalue of matrix a, and I is an identity matrix in the same size as a. We consider a 2-by-2 matrix for illustration:

$$a = \begin{pmatrix} a_{1,1} & a_{1,2} \\ a_{2,1} & a_{2,2} \end{pmatrix} \qquad (1.56)$$

Then, $a - \lambda I$ becomes more explicitly

$$a - \lambda I = \begin{pmatrix} a_{1,1} & a_{1,2} \\ a_{2,1} & a_{2,2} \end{pmatrix} - \lambda \begin{pmatrix} 1 & 0 \\ 0 & 1 \end{pmatrix} = \begin{pmatrix} a_{1,1} - \lambda & a_{1,2} \\ a_{2,1} & a_{2,2} - \lambda \end{pmatrix} \qquad (1.57)$$

Therefore, Eq.(1.55) becomes

$$\det(a - \lambda I) = \det \begin{pmatrix} a_{1,1} - \lambda & a_{1,2} \\ a_{2,1} & a_{2,2} - \lambda \end{pmatrix} = 0 \qquad (1.58)$$

or equivalently

$$(a_{1,1} - \lambda)(a_{2,2} - \lambda) - a_{1,2}a_{2,1} = 0 \qquad (1.59)$$

which is a quadratic polynomial of λ and has two roots. These roots are eigenvalues of matrix a.

Likewise, if matrix a is a 3-by-3 matrix, it has 3 roots, namely 3 eigenvalues. With Octave/Matlab, eigenvalues of matrix a are computed by **eig(a)** command. For illustration, eigenvalues of

$$a = \begin{pmatrix} 3 & 2 & 0 \\ 1 & 2 & 5 \\ 9 & 1 & 2 \end{pmatrix} \qquad (1.60)$$

is computed as follows:

>>a=[3 2 0; 1 2 5; 9 1 2], eig(a)

ans=
   7.31961 + 0.00000i
  -0.15980 + 3.36361i
  -0.15980 - 3.36361i

Thus the matrix a has one real eigenvalue and two complex conjugate eigenvalues.

Extending what we have seen for the 2-by-2 matrix, $\det(a - \lambda I)$ for any matrix a of order n is a characteristic polynomial of order n, which we denote by

$$f(\lambda) = \det(a - \lambda I) \tag{1.61}$$

Therefore, eigenvalues of a matrix are roots of its characteristic polynomial. To write down the characteristic polynomial by hand calculation of the determinant is time consuming, but very simple if Octave/Matlab is used. Indeed, by the command **c=poly(a)**, c becomes an array of the power coefficients of the characteristic polynomial as shown next:

```
>> a=[3 2 0; 1 2 5; 9 1 2]; c=poly(a)
c =
   1.0000  -7.0000   9.0000  -83.0000
```

Then, roots of the polynomial can be computed by **roots** command:

```
>> roots(c)
ans =
   7.31961 + 0.00000i
  -0.15980 + 3.36361i
  -0.15980 - 3.36361i
```

The foregoing roots agree exactly with the eigenvalues computed using **eig(a)**. Indeed, **roots(poly(a))** is equivalent to **eig(a)**.

We note here that all eigenvalues are real if the matrix is symmetric. We show one example with Octave/Matlab calculations:

```
>>a=[3 2 5; 2 2 5; 5 5 2], eig(a)
a =
   3  2  5
   2  2  5
   5  5  2
ans =
  -3.95394
   0.50847
  10.44547
```

1.8 Limitations of Octave/Matlab in linear algebra

Solving linear equations with Octave/Matlab becomes increasingly slower as the number of unknowns increases.

Such a linear set of equations with a large number of unknowns occurs particularly when an elliptic partial differential equations are approximated by a finite difference method. A modest problem of a two-dimensional elliptic partial differential equation would have 100x100 unknowns, or equivalently 10,000 unknowns. Its coefficient matrix becomes a 10,000x10,000 matrix. Such a problem cannot be solved with Octave/Matlab on an ordinary desktop computer.

A better method of solving such a huge linear set of equations is an iterative method. Although iterative methods cannot solve every linear set of equations, the linear set of equations derived for certain classes of elliptic partial differential equations have such mathematical properties that assure convergence of iterative schemes.

Nonetheless, iterative methods may be practiced with Octave/Matlab. More details will be written in a book to be published by the author late in 2016.

Problems for Chapter 1

[1] Matrix a and vector b are given respectively by $a = $ [1 2 3; 0 1 4; 3 0 2], $b = $ [3 5 1]'. Calculate ab by hand calculation, and verify the results by Octave or Matlab.

[2] Two matrices a and b are given respectively by $a = $ [1 2 3; 0 1 4; 3 0 2], $b = $ [1 1 1; 2 -1 3; 3 2 -2]. Calculate ab by hand calculations, and verify by Octave or Matlab.

[3] Solve the following linear equations by Octave or Matlab:

(a)
$$x_1 - x_2 = 5$$
$$3x_1 + 2x_2 = 7$$

(b)
$$3x_1 + x_2 + 2x_3 = 2$$
$$-x_1 + x_2 + 4x_3 = 0$$
$$3x_1 + x_2 + 2x_3 = 5$$

[4] The following matrices are diagonal matrices. Calculate their inverse by hand calculation.

$$a = \begin{pmatrix} 4 & 0 \\ 0 & 5 \end{pmatrix}, \quad b = \begin{pmatrix} 3 & 0 & 0 \\ 0 & 1 & 0 \\ 0 & 0 & 2 \end{pmatrix}$$

[5] Is the following set of linear equations ill-conditioned or not? Compute the condition number. Solve the equation by Octave/Matlab, and if possible by Fortran or C in single precision.

$$3.2406x_1 + 2.9155x_2 = 7.00000$$
$$-9.7128x_1 - 8.7464x_2 = -2.97067$$

[6] Can you believe the solution of the equation in Problem [5] accurate? State the reasons.

[7] Calculate the product of two matrices namely aa^{-1} where a is a 10x10 Hilbert matrix. Discuss what you observe.

[8] Matrix a is given by **a = [3 4 -2; 3 -1 1; 2 0 5]**. Find the eigenvalues by the **eig** command. Also determine the characteristic polynomial and find the roots by **roots** command.

[9] The following set of linear equations is an eigenvalue problem:
$$x_1 + 2x_2 + 2x_3 = \lambda x_1$$
$$-x_1 - x_2 + 4x_3 = \lambda x_2$$
$$3x_1 + 5x_2 + x_3 = \lambda x_3$$
Find the characteristic polynomial and the roots by **poly** and **roots** commands. Verify the results by the **eig** command.

[10] Calculate the determinant of a and b in [4] by hand calculations.

[11] Calculate the eigenvalues of a and b in [4] by hand calculation.

[12] Calculate the eigenvalues of a^{-1} and b^{-1} for a and b in [4] by hand calculation.

[13] Calculate the condition number of a and b in [4] by hand calculation.

Chapter 2
Polynomials and Polynomial Interpolation

Polynomials are foundation of many numerical methods such as numerical integration, interpolation, and differentiation methods. That is why understanding polynomials is so important. This chapter describes both polynomials and polynomial interpolation.

Octave/Matlab commands used in this chapter:
interp2(X, Y, F, x1, y1, method): 2-dimensional interpolation, X, Y are arrays of the coordinates of the data points, F is function table, $x1$ and $y2$ are input for which the interpolated values are to be evaluated, and method is the method of interpolation
roots(c): roots of polynomial c
poly(r): power coefficients of normalized polynomial when all roots are given by array r
polyder(c): derivative of polynomial c
polyfit(x, y, n): power coefficients of the polynomial of order n fitted to the data points, in the array form, x and y
polyval(c, x): values of polynomial c at x

2.1 Different forms of polynomials

A polynomial can be equivalently expressed in different forms. The polynomial written in one form can be converted to another. Some of the transformations may need help of Octave/Matlab, but others may be done using the power coefficients, or simply writing down using given information.

Power series form
Polynomials are functions that consist of powers of an independent variable, x, such as x, x^2, x^3 ...

The simplest polynomial is the linear function,

$$y = c_1 x + c_2 \qquad (2.1)$$

---

where c_1 and c_2 are constants named power coefficients. If plotted on a graph the linear equation is a line as illustrated in Figure 2.1 where $y = 0.5x - 1$ with $c_1 = 0.5$ and $c_2 = -1$ are assumed for illustration purpose.

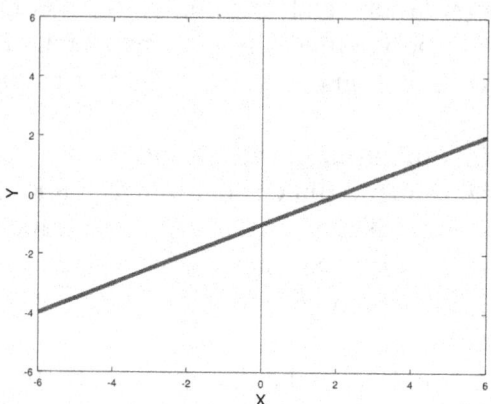

Figure 2.1 Graphic plot of a linear equation

```
% C2_1.m     Plots Figure 2.1
clear, clf
x=-6:1:6; y = 0.5*x - 1;
plot(x,y,'linewidth', 4); hold on
plot([-6,6], [0,0],'b', [0,0],[-6,6],'b');
xlabel('X', 'fontsize', 16)
ylabel('Y', 'fontsize', 16)
axis([-6, 6, -6,6])
```

The next simplest polynomial is the quadratic polynomial written as

$$y = c_1x^2 + c_2x + c_3 \qquad (2.2)$$

An example of the quadratic polynomial,

$$y = 3x^2 - 2x - 7 \qquad (2.3)$$

is plotted in Figure 2.2. Roots of this polynomial are 1.8968 and 1.2301

Every quadratic polynomial has two roots, which can be two real roots, one double root, or two complex conjugate roots. The real roots, including double root, can be identified on a graph. For the foregoing quadratic polynomial, the two real roots are marked by dark circles, which are at intersection of the polynomial and the x-axis. In case the roots are complex conjugate, the polynomial does not intersect the x-axis.

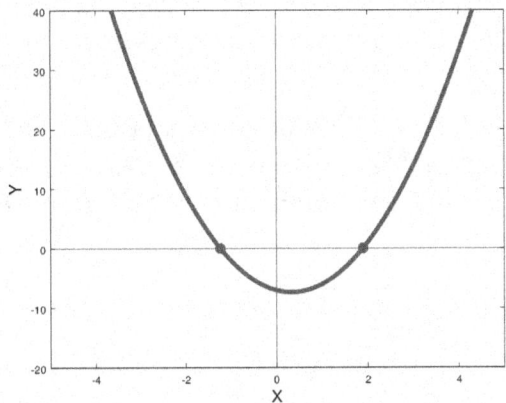

Figure 2.2 Graphic illustration of a quadratic polynomial

```
%C2_2.m    Plots Fig 2.2
clear, clf, hold on;
x = -5:.1:5; c= [3, -2, -7];
y=polyval(c,x);
r=roots(c);
plot(x,y, 'linewidth',4)
plot( [-10,10], [0,0],'b', [0,0],[-20,100],'b'); hold on
xlabel('X', 'fontsize', 16)
ylabel('Y', 'fontsize', 16)
for k=1:length(r)
plot(r(k), 0,'o', 'linewidth', 4)
end %for
axis([-5 5 -20 40])
```

A cubic polynomial as well as a 4th order polynomial are, respectively,

$$y = c_1 x^3 + c_2 x^2 + c_3 x + c_4 \qquad (2.4)$$

$$y = c_1x^4 + c_2x^3 + c_3x^2 + c_4x + c_5 \qquad (2.5)$$

In more general, an nth order polynomial is

$$y = c_1x^n + c_2x^{n-1} + .. + c_{n-1}x^2 + c_nx + c_{n+1} \qquad (2.6)$$

where c_1, c_2, ... c_{n+1} are power coefficients, which are all real numbers. In Octave/Matlab, a polynomial is represented by an array c, where $c=[c_1 \ c_2 \ ... \ c_{n+1}]$. The highest order of x in this polynomial is n, and the polynomial is said to be a polynomial of order n, or n^{th} order polynomial. A polynomial of order n has a maximum of $n+1$ coefficients, some of which may be zero except the leading coefficient.

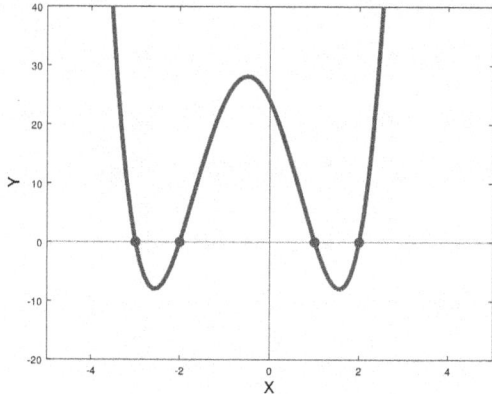

Figure 2.3   Fourth-order polynomial

```
%C2_3.m    Plots Fig 2.3
clear, clf, hold on;
x = -5:.1:5; c= [2  4 -14 -16 24];
y=polyval(c,x);
r=roots(c);
plot(x,y, 'linewidth',4)
plot( [-10,10], [0,0],'b', [0,0],[-20,100],'b'); hold on
xlabel('X', 'fontsize', 16)
ylabel('Y', 'fontsize', 16)
for k=1:length(r)
plot(r(k), 0,'o', 'linewidth', 4)
```

```
end %for
axis([-5 5 -20 40])
```

The roots of a polynomial can be obtained by the **roots** command. Considering a sample polynomial

$$y = 2x^4 + 4x^3 - 14x^2 - 16x + 24 \qquad (2.7)$$

its roots are computed as

```
>> c=[2  4 -14 -16 24]; roots(c)
ans =
 -3.0000
 -2.0000
  2.0000
  1.0000
```

If a polynomial has multiple (or repeated) roots or some roots are very close to each other, accuracy of **roots** can be poor. See Appendix B for more details.

The derivative of a polynomial in the power series form of Eq.(2.6) may be obtained by **polyder(c)** command.

Clustered form
The clustered form is given by

$$y = \left(c_1 x + c_2\right) \qquad (2.8)$$

$$y = \left(\left(c_1 x + c_2\right)x + c_3\right) \qquad (2.9)$$

...

$$y = ((..((c_1 x + c_2)x + c_3)x..+ c_n)x + c_{n+1}) \qquad (2.10)$$

Factorized form
A polynomial of order n has n roots, some of which may be complex values and some may be multiple (or repeated). In terms of the roots, r_1, r_2, ... r_n, the polynomial can be expressed as

$$y = c_1(x-r_1)(x - r_2) \dots (x - r_n) \tag{2.11}$$

Normalized polynomial

A polynomial that is divided through by its leading coefficient, whether the form is a power series, clustered or factorized, is said to be **normalized**. For example, the following polynomials are normalized:

$$y = (x-1)(x-2)(x+2)(x+3) \tag{2.12}$$

$$y = x^4 + 2x^3 - 7x^2 - 8x + 12 \tag{2.13}$$

Lagrange form

An nth order polynomial can be determined if $n+1$ data points that the polynomial passes through are given. Denoting these $n+1$ data points by (x_i, y_i), $i = 1, 2, ..n+1$, the polynomial in the Lagrange form is

$$y = \sum_{i=1}^{n+1} g_i(x) y_i \tag{2.14}$$

with

$$g_i(x) = \prod_{j=1, j\neq i}^{n+1} \frac{(x - x_j)}{(x_i - x_j)} \tag{2.15}$$

Here, each of $g_i(x)$ is a polynomial of order n, which becomes zero at every data point x_j, except at x_i where $g_i(x_i) = 1$. Therefore, Eq. (2.14) is a polynomial of order n and passes through every data point (x_i, y_i), $i = 1, 2,.., n+1$. For a few values of n, Eq.(2.14) is written more explicitly below:

$$y = \frac{x - x_2}{x_1 - x_2} y_1 + \frac{x - x_1}{x_2 - x_1} y_2, \quad (n=1) \tag{2.16}$$

$$y = \frac{(x - x_2)(x - x_3)}{(x_1 - x_2)(x_1 - x_3)} y_1 + \frac{(x - x_1)(x - x_3)}{(x_2 - x_1)(x_2 - x_3)} y_2$$

$$+ \frac{(x - x_1)(x - x_2)}{(x_3 - x_1)(x_3 - x_2)} y_3, \quad (n=2) \tag{2.17}$$

2.2 How to define a unique polynomial

The main question in this section is how to define a polynomial of order n uniquely. There are a few different ways to do this.

All of $n+1$ power coefficients are specified
In this case, the power series form or clustered form of the polynomial can be written immediately.

A set of $n+1$ data points that the polynomial passes through are known
Why this is true is explained first. Let us write the known data points by (x_i, y_i), $i = 1, 2, ..$ $n+1$. Writing Eq.(2.6) for each data point yields $n+1$ equations:

$$c_1 x_i^n + c_2 x_i^{n-1} + .. + c_{n-1} x_i^2 + c_n x_i + c_{n+1} = y_i \tag{2.18}$$

where $i = 1, 2, ..n+1$, and the left side and right side have been switched. Now we have $n+1$ equations for $n+1$ unknowns c_i, $i = 1, 2, ..n+1$. So we can solve Eq.(2.18), which yields a unique set of power coefficients.

With Octave/Matlab, no human labor of solving the linear equation is necessary. Using the **polyfit** command, the power coefficients are determined. For example, assume three data points are (0, 2), (1, 0) and (3, 1), or equivalently x=[0 1 3], y=[2 0 1], then

41

```
>> x=[0 1 3]; y=[2 0 1]; c=polyfit(x,y,length(x)-1)
c=0.83333 -2.83333  2.00000
```

which means the polynomial is

$$y = 0.83333x^2 - 2.83333x + 2.00000 \qquad (2.19)$$

We also note that a polynomial in the Lagrange form is determined uniquely when $n+1$ data, (x_i, y_i), are specified.

All roots of the polynomial and one data point that the polynomial passes through are specified

A un-normalized polynomial of order n has $n+1$ coefficients. In order to determine the $n+1$ coefficients, $n+1$ pieces of information are necessary. When n roots of an nth order polynomial are used to determine the normalized form, in which the leading coefficient is unity, we are using only n pieces of information. This normalized polynomial is not a unique polynomial because a normalized polynomial times any constant has the same roots. In order to convert the normalized form to a unique un-normalized form, one more piece of information is necessary.

Suppose we write the coefficients of a normalized polynomial in an array form by d, the leading element of which is unity. Now, if the unique polynomial is to pass through a data point, (x_0, y_0), as the $(n+1)^{th}$ piece of information, then the normalized polynomial times a constant a must be satisfied

$$a(x_0^n + d_2 x_0^{n-1} + .. + d_{n-1} x_0^2 + d_n x_0 + d_{n+1}) = y_0 \qquad (2.20)$$

So the constant a can be determined by solving Eq.(2.20) because all other terms are known. To rewrite the foregoing equation to a standard power series form,

$$c_1 x_0^n + c_2 x_0^{n-1} + .. + c_{n-1} x_0^2 + c_n x_0 + c_{n+1} = y_0 \qquad (2.21)$$

---

the power coefficients are

$$c_1 = a \qquad\qquad (2.22)$$

and

$$c_i = c_1 d_i , \quad i = 2,.. \, n+1 \qquad\qquad (2.23)$$

2.3 Evaluation of a polynomial

In the prior section we learned how to write a unique polynomial. All polynomials can be then expressed by an array, c, of the power coefficients. The value of the polynomial can be evaluated by the **polyval** command.

Example
A polynomial

$$y = x^4 + 2x^3 - 7x^2 - 8x + 12$$

can be evaluated for $x = -1, 0, 2$ and 5 as follows:

```
>>x=[-1, 0, 2, 5]; c=[1, 2, -7, -8, 12]; y=polyval(c,x)
y =
   12   12   0  672
```

which are values of the polynomial for $x= -1, 0, 2$ and 5, respectively.

2.4 Polynomial interpolation

Polynomial interpolation is to fit a polynomial to a certain number of data points sampled from a function, whether it is a known function or unknown function. It is important in developing a large number of algorithms in numerical methods.

There are two different ways of fitting polynomial(s) to a given data set. The first is to use a single polynomial for the entire data set, and the second is that the entire domain is partitioned into a number of subintervals and a low order polynomial is fitted in each subsubinterval.

2.4.1 Fitting a single polynomial in the entire domain

This approach is used when a relatively small number of data points are used in the entire domain. The simplest interpolation formula is the linear interpolation. A linear interpolation is fitted to two data points, (x_1, y_1) and (x_2, y_2). As an example, we illustrate a linear interpolation fitted to two data points sampled from $\sin(x)$ in $0 \le x \le \pi/4$. The reason for sampling the data points from a known function here is that, by knowing the exact function, error of the interpolation can be evaluated easily. This will help us understand the nature of errors in polynomial interpolations.

The two points of the sine function in this case are (0, 0) and $(\pi/4, \sin(\pi/4))$ where the first number in the parentheses is the value of x and the second the value of y for a point. The coeff-icients of the linear equation,

$$y = c_1 x + c_2 \tag{2.24}$$

can be determined by solving the following two equations

$$y_1 = \sin(0) = c_1 * 0 + c_2 \tag{2.25}$$

$$y_2 = \sin(\pi/4) = c_1 * \pi/4 + c_2 \tag{2.26}$$

where the first equation is for the first point (0, 0), and the second for the second point, $(\pi/4, \sin(\pi/4))$. From the first equation we get immediately $c_2 = 0$, and then from the second equation

$$\sin(\pi/4) = c_1 * \pi/4 \tag{2.27}$$

so,

$$c_1 = \sin(\pi/4)/(\pi/4) = 0.90032 \qquad (2.28)$$

Thus, the interpolation formula is

$$y = 0.90032x \qquad (2.29)$$

The interpolation polynomial like Eq.(2.29) can be obtained with Octave/Matlab by using the **polyfit** command as follows:

```
>> t = [0, pi/4] ; c=polyfit(t, sin(t), 1)
c =
   9.0032e-001  7.8505e-017
```

The second element in the above c is same as zero. Therefore, the answer agrees with Eq. (2.29).

In order to see how accurate this interpolation formula is, we plot it and compare to sin(x), as shown in Fig. 2.4, where the dotted curve is the exact function sin(x). Figure 2.4 indicates that the error tends to become highest about the middle of the interpolation range.

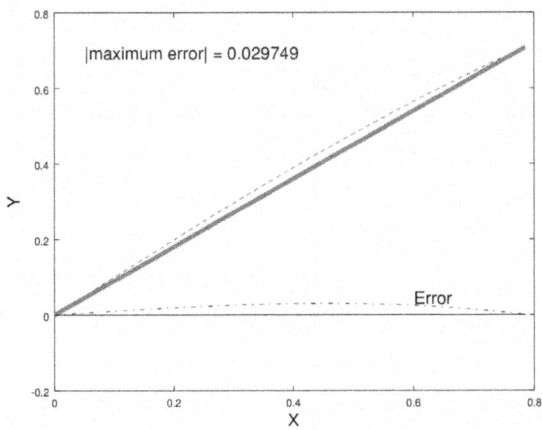

Figure 2.4 Plot of a linear interpolation and its error

%C2_4.m   Plots Fig 2.4clear,clf

```
x = 0:pi/4/20:pi/4;
f = sin(x);
xp=[0,pi/4];
c=polyfit(xp, sin(xp),1)
g=polyval(c,x)
plot(x,f ,'--',x ,g, 'linewidth',4); hold on
plot(x, (f - g), x, 0*x, 'k')
maxerror=max(abs(f-g));
me= ['|maximum error| = ', num2str(maxerror) ]
text(0.05, 0.7, me ,'fontsize', 16)
text(0.6, max(abs(f-g))+0.02, 'Error', 'fontsize', 16)
xlabel('X', 'fontsize', 16)
ylabel('Y', 'fontsize', 16)
```

Now we use the quadratic polynomial to approximate the sine function in the same range. The power coefficients of the quadratic polynomial are determined by Octave/Matlab as follows:

>>t = [0, pi/8, pi/4] ; c=polyfit(t, sin(t), 2)
c =
 -0.18890   1.04867 -0.00000

So, the quadratic polynomial interpolation is

$$y = -0.18890x^2 + 1.04867x$$

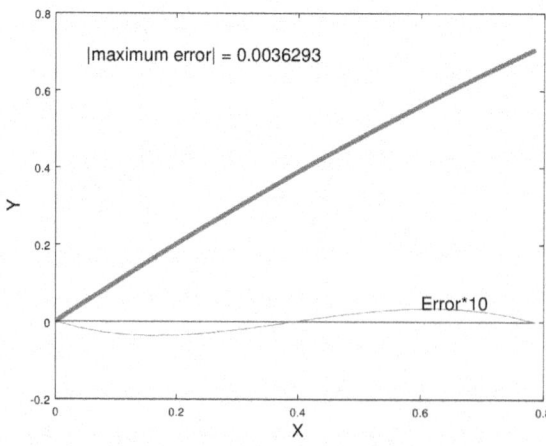

Figure 2.5 Plot of a quadratic polynomial interpolation and its error distribution

In order to illustrate some important aspect of polynomial interpolation, we increase the number of data points from 3 to 5 for the sine function and for the same interval from 0 to $\pi/4$:

```
>>t = [0,  pi/8/2, pi/4] ; c=polyfit(t, sin(t), 4)
c =
 -1.1501e-001  -1.1037e-001  4.3125e-002  9.9024e-001  -1.1102e-016
```

The order of polynomial is now 4 and the polynomial in power series form is given by

$$y = -0.11501x^4 - 0.11037x^3 + 0.043125x^2 + 0.99024x \qquad (2.30)$$

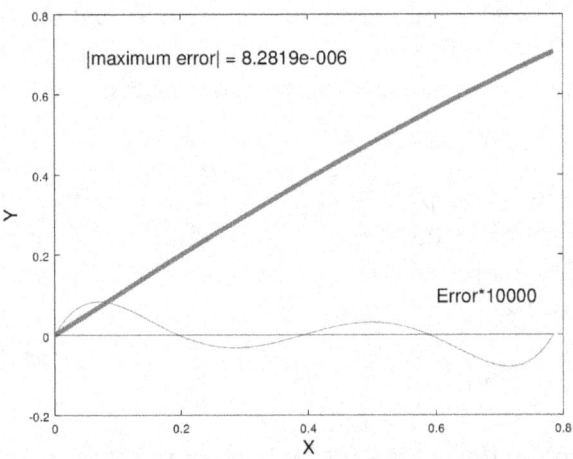

Figure 2.6 A 4th order polynomial interpolation
and its error distribution

```
%C2_5.m    Plots Fig 2.6
clear,clf
t = [0:pi/8/2:pi/4] ;
c=polyfit(t, sin(t), 4);
x = 0:pi/4/20/4:pi/4;
f = sin(x);
g =  polyval(c, x);
clf, plot(x,f ,'--',x ,g, 'linewidth',4); hold on
plot( x, (f - g)*10000, x, 0*x, 'k')
maxerror=max(abs(f-g));
```

```
me= ['|maximum error| = ', num2str(maxerror) ]
text(0.05, 0.7, me ,'fontsize', 16)
text(0.55, max(abs(f-g)*10000)+0.02, 'Error*10000', 'fontsize', 16)
xlabel('X', 'fontsize', 16)
ylabel('Y', 'fontsize', 16)
```

Since the fitted 4th order polynomial became so close to the exact function, $\sin(x)$, they cannot be distinguished from each other in Figure 2.6. The error times 10000 is plotted also. The error is significantly smaller than the quadratic interpolation polynomial illustrated in Figure 2.5. As the order of polynomial increases, the error of interpolation decreases. However, the error plotted in Figure 2.6 shows one important trend, that is, the peak of fluctuating errors tends to be higher toward both ends of the interpolation range. This trend intensifies as the order of polynomial increases. In general, the accuracy of interpolation with equispaced data points is the best in the mid-section of the interpolation range, but error increases toward the edges.

Understanding this trend of polynomial fitting will become helpful when polynomial interpolations are applied to derive other numerical methods. For example, central difference approximations are always more accurate than one-sided difference approximations, because the former uses the center of an interpolation polynomial while the latter uses off-centered position of an interpolation polynomial.

2.4.2 Polynomial interpolation with Chebyshev points

The trend of increasing error toward the ends of the interpolation range may be eradicated by using the Chebyshev points, which are non-uniformly distributed data points determined by a Chbyshev polynomial. If the order of the interpolation polynomial is n, $n+1$ Chebyshev points are necessary, which are given by

$$x_m = 0.5\left[(b-a)\cos\frac{(k+0.5-m)\pi}{k} + a + b \right]$$

$$m = 1, 2, \ldots k \qquad\qquad (2.31)$$

where $k = n+1$ is the number of data points, a and b are the x-values of the end points of the interpolation range.

Consider approximating $\sin(x)$ in $0 \le x \le \pi/4$ by a polynomial of order 4. In this case $a = 0$, $b = \pi/4$, and $n = 4$. The Octave/Matlab calculation of the Chebyshev points is as follows.

```
>> k=5; m=1:k; a = 0; b = pi/4;
>>t = 0.5*((b-a)*cos((k+0.5-m)*pi/k) + a + b)
t =
    0.019220  0.161876  0.392699  0.623522  0.766178
```

The interpolation polynomial with the 5 Chebyshev points is plotted in Figure 2.7, in which the maximum error and the distribution of errors are also printed. Although errors fluctuate, all of their local peaks are nearly in the same magnitudes, and importantly the maximum error with Chebyshev points becomes approximately a half of that with equispaced points.

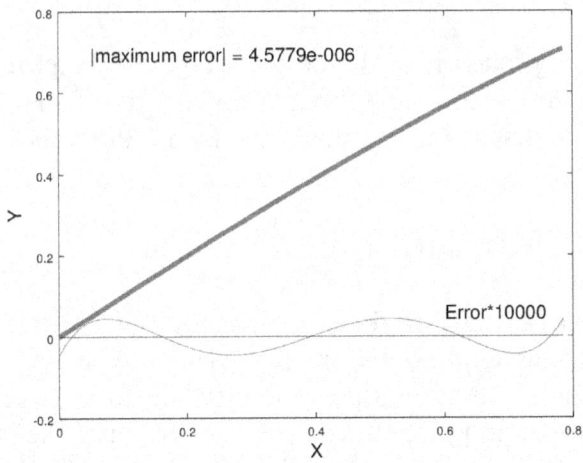

Figure 2.7   The interpolation polynomial
with the 5 Chebyshev points

```
%C2_6.m    Plots Fig 2.7
clear, clf
k=5;m=1:k; a = 0; b = pi/4;
t = 0.5*((b-a)*cos((k+0.5-m)*pi/k) + a + b);
```

```
c=polyfit(t, sin(t), 4);
x = 0:pi/4/20/4:pi/4;
f = sin(x);
g = polyval(c, x);
clf, plot(x,f ,'--',x ,g, 'linewidth',4); hold on
plot(x, (f - g)*10000, x, 0*x, 'k')
maxerror=max(abs(f-g));
me= ['|maximum error| = ', num2str(maxerror) ]
text(0.05, 0.7, me ,'fontsize', 16)
text(0.55, max(abs(f-g)*10000)+0.02, 'Error*10000', 'fontsize', 16)
xlabel('X', 'fontsize', 16)
ylabel('Y', 'fontsize', 16)
```

2.4.3 Effect of interpolation range size

In the previous section, the effects of increasing number of interpolation points and changing the location of data points are discussed. However, it is important to note that the size of the interpolation range has a significant impact on the accuracy of interpolation. In other words, as the size of interpolation range decreases, the accuracy becomes better, and conversely, the opposite is true.

Thus, a wise approach is to divide the whole range into subintervals and then use a low order polynomial interpolation for each subinterval as discussed in more details in the following section.

2.5 Piecewise polynomial interpolations

2.5.1 Piecewise linear interpolation

In the piecewise linear approximation, the whole range of interpolation is subdivided into multiple subintervals, and in each subinterval, a linear interpolation is applied. Because the data points are located at the boundaries of subintervals and shared between the two adjoining subintervals, the piecewise linear interpolation becomes continuous throughout the whole domain.

In the following examples, we divide the range of $0 \leq x \leq \pi / 4$ into 4 subintervals as well as 8 subintervals. The results are shown in Figures 2.8 and 2.9 respectively. As printed in the figures, the maximum error in case of 4 subintervals is 0.003056, and that of 8 subintervals is 0.00080891, the latter of which is approximately ¼ of the former.

```
%C2_7.m    Plots Fig 2.8
clear, clf; hold on
t = [0:pi/8/2:pi/4] ;
L=length(t);
xt=0; yt=0;
for m=1:L-1
a1 = t(m); a2= t(m+1);
b1=sin(a1); b2=sin(a2);
plot([a1,a2],[b1 b2], 'linewidth',4)
x1=a1:0.001:a2;
y1=sin(x1);
plot(x1,y1, '--')
c=polyfit([a1,a2],[b1 b2], 1);
g=polyval(c, x1);
plot(x1,10*(sin(x1)-g), x1, x1*0, 'k')
xt=[xt, x1]; yt=[yt, sin(x1)-g];
end
maxerror=max(abs(yt));
me= ['|maximum error| = ', num2str(maxerror) ]
text(0.05, 0.7, me ,'fontsize', 16)
text(0.6, maxerror*10+0.02, 'Error*10', 'fontsize', 16)
xlabel('X', 'fontsize', 16)
ylabel('Y', 'fontsize', 16)
```

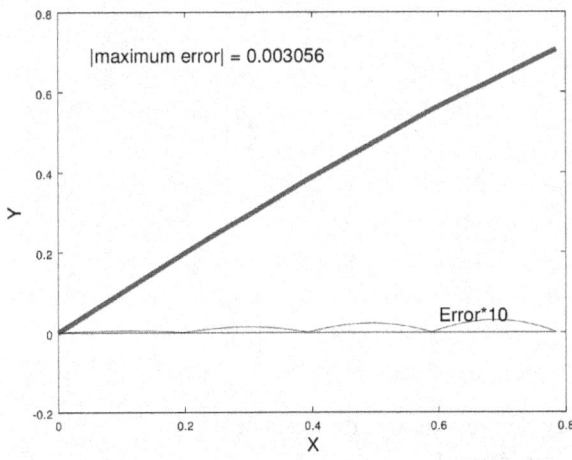

Figure 2.8 Piecewise linear interpolation with 4 subintervals

```
%C2_8.m     Plots Fig 2.9
clear; clf; hold on
t = [0:pi/8/4:pi/4] ;
L=length(t);
xt=0; yt=0;
for m=1:L-1
a1 = t(m); a2= t(m+1);
b1=sin(a1); b2=sin(a2);
plot([a1,a2],[b1 b2], 'linewidth',4)
x1=a1:0.001:a2;
y1=sin(x1);
plot(x1,y1, '--')
c=polyfit([a1,a2],[b1 b2], 1);
g=polyval(c, x1);
plot(x1,20*(sin(x1)-g), x1, x1*0, 'k')
xt=[xt, x1]; yt=[yt, sin(x1)-g];
end
maxerror=max(abs(yt));
me=  ['|maximum error| = ', num2str(maxerror) ]
text(0.05, 0.7, me ,'fontsize', 16)
text(0.6, maxerror*20+0.02, 'Error*20', 'fontsize', 16)
xlabel('X', 'fontsize', 16)
ylabel('Y', 'fontsize', 16)
```

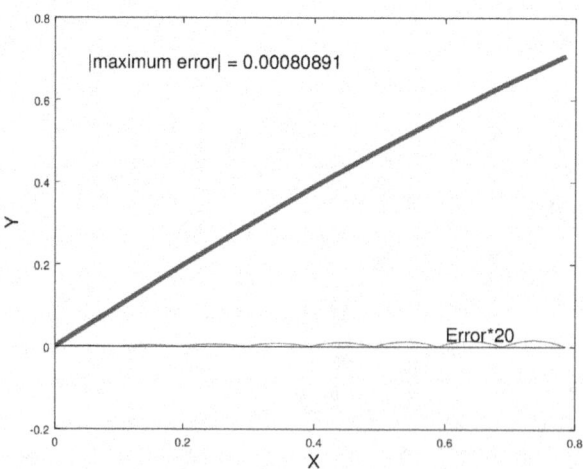

Figure 2.9 Piecewise linear interpolation with 8 subintervals

2.5.2 Piecewise higher order polynomial interpolations

The concept of piecewise linear interpolation can be extended to a higher order polynomial in each subinterval. With higher order polynomials, more sophisticated and interesting applications become possible. A very brief overview is given next.

Cubic Hermite polynomial fitting to both functional data and the first derivative: When a cubic polynomial is used in each subinterval, four conditions can be imposed on the polynomial for each subinterval. Therefore, in addition to a functional value, its first derivative may be satisfied at each end of the subinterval. The second derivative of cubic Hermite polynomials is not continuous across the boundary between two consecutive subintervals.

C-spline function: it uses cubic polynomial in each subinterval. It passes through the functional data given at the boundaries of the subintervals. However, both first derivative and second derivative of the cubic polynomials are required to be continuous at each boundary of two consecutive subintervals. However, unlike the Hermite cubic polynomial, the first derivatives cannot be prescribed at the ends of each subintervals.

B-spline function: b-spiline function is based on cubic polynomial in each subinterval and is a sibling of the c-spline function, but has different characters. The given data points are used as control points, but the b-spline function does not necessarily pass through the control points. The b-spline function is often used to create smooth and eye-pleasing curves.

More details of the c-spline and b-spline functions may be found in Reference [Nakamura, 2002].

2.6 Two-dimensional interpolation

2.6.1 Bilinear interpolation

Data in a two-dimensional function table can be interpolated using linear interpolation twice. The two-dimensional function table, as illustrated in Table 2.1, is an array of functional values $f_{i,j}=f(x_i, y_j)$ on a rectangular mesh, (x_i, y_j).

Table 2.1   A sample data table

i		1	2	3	4
j	y\x	0.0	0.2	0.4	0.6
1	0.0	0.90000	0.90241	0.90952	0.92100
2	0.4	0.90789	0.91861	0.93325	0.95111
3	0.8	0.93033	0.94766	0.96752	0.98894
4	1.2	0.96376	0.98498	1.00691	1.02852

```
clear    % Table 2.1  Sample data table
x=0:0.2:0.6;
y=0:0.4:1.2;
[X,Y]=meshgrid(x,y);
f=1-0.1*cos(X*1.1+Y)
```

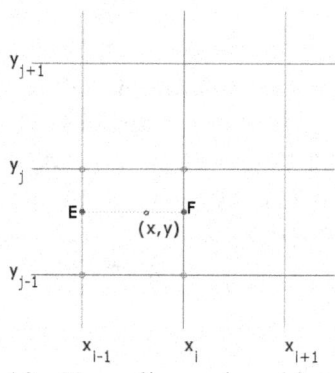

Figure 2.10   Two-dimensional interpolation

Suppose we are to find an approximate functional value at a point (x, y) located in a rectangular domain defined by $x_{i-1} \le x \le x_i$ and $y_{j-1} \le y \le y_j$, as shown in Figure 2.10. By linear interpolation in the y-direction, the values at E and F are found, respectively, as

$$f_E = \frac{y_j - y}{y_j - y_{j-1}} f_{i-1,j-1} + \frac{y - y_{j-1}}{y_j - y_{j-1}} f_{i-1,j} \tag{2.32}$$

$$f_F = \frac{y_j - y}{y_j - y_{j-1}} f_{i,j-1} + \frac{y - y_{j-1}}{y_j - y_{j-1}} f_{i,j} \tag{2.33}$$

The linear interpolation in the x-direction using of f_E and f_F yields the bilinear interpolation:

$$g(x,y) = \frac{x_i - x}{x_i - x_{i-1}} f_E + \frac{x - x_{i-1}}{x_i - x_{i-1}} f_F \qquad (2.34$$

In Octave/Matlab, bilinear interpolation is performed by **interp2**, use of which is illustrated in the following subsection.

2.6.2   Double Lagrange interpolation

For a higher accuracy of two-dimensional interpolation, the double Lagrange interpolation may be a choice, which is to apply the Lagrange interpolation method twice in two-dimensions. Therefore, the interpolation uses all the data points in the table. Suppose the function table has I columns for x points and J rows for y points. Using the notations for the coordinates of the points, the double Lagrange interpolation is written as

$$g(x,y) = \sum_{i=1}^{I} \sum_{j=1}^{J} u_i(x) v_j(y) f_{i,j} \qquad (2.35)$$

where

$$u_i(x) = \prod_{k=1, k \neq i}^{I} \frac{x - x_k}{x_i - x_k} \qquad (2.36)$$

$$v_j(x) = \prod_{k=1, k \neq j}^{J} \frac{y - y_k}{y_j - y_k} \qquad (2.37)$$

When $I = J = 2$, Eq.(2.35) reduces to the bilinear interpolation, Eq.(2.32). The Chebyshev points may be used to select x_i, and y_j.

Octave/Matlab does not provide a command for double Lagrange interpolation but **interp2** command has options for linear, cubic and spline ("linear" is the same as "bilinear"):

>>f=interp2(X, Y, F, XI, YI, Method)

The last input "Method" specifies the method of interpolation. Choices include 'linear' (default), 'cubic' and 'spline'. For more details use >>help interp2. In the following example, linear, cubic and spline are all used (on Octave) for comparison.

```
clear
x=0:0.2:0.6;
y=0:0.4:1.2;
[X,Y]=meshgrid(x,y);
F=[
    0.90000  0.90241  0.90952  0.92100;
    0.90789  0.91861  0.93325  0.95111;
    0.93033  0.94766  0.96752  0.98894;
    0.96376  0.98498  1.00691  1.02852;
    ]
f_linear=interp2(X, Y, F, 0.1, 0.2, 'linear')
f_cubic=interp2(X, Y, F, 0.1, 0.2, 'cubic')
f_spline=interp2(X, F, F, 0.1, 0.2, 'spline')
```
Results (Octave)
```
f_linear =  0.90723    Error= -0.0024609
f_cubic =  0.90538     Error= -6.1028e-004
f_spline =  0.90468    Error= 8.8245e-005
```

Here the calculation of the errors was possible because the data table was generated by the script after Table 2.1, and the function used is known.

Problems for Chapter 2
(Solve all the problems using Octave/Matlab but not by hand calculations unless otherwise instructed.)

[1] Determine the coefficients of the linear function passing through $y = -5$ on the y-axis and $x = 2$ on the x-axis.

[2] Determine the power coefficients of the quadratic polynomial that passes though the following three points for each case:
   (a) (-1,-3), (0,1), (2,-1)
   (b) (-3,6), (0,-2), (2,1)

[3] Transform the following polynomial to a cluster form, normalized form and a factorized form:
$$y = 2x^3 - 4x^2 + x + 3$$

[4] Evaluate the values of
$$y = x^3 - 7x^2 + 4x - 5$$
for $x = -5, -2, +1, +3, +5$. Hint: use **polyval**.

[5] Find the roots of the following polynomials
(a) $y = 2x^2 - 4x - 6$
(b) $y = x^2 - 2x - 3$
(c) $y = x^3 + x^2 - 9x - 9$
(d) $y = x^5 - 2x^3 + 4x - 3$

[7] What is the order of the unique polynomial that passes though the following points?
(a) (0, 2), (1, -1), (2, 0)
(b) (1, 0), (3, 0), (5, 0), (7, 1)
(c) (1, 0), (3, 1), (5, 0), (7, 0)

[8] Determine the polynomial in power series form passing through the points for each case given in Problem [7].

[9] Plot the polynomial for each problem in [8].

[10] Write a script that reads the data points as input and plot the unique polynomial passing through the given data points. The script must work for any polynomial in power series. The figure plotted must include axes notations X or Y printed for each axis.

[11] A polynomial is known to have two roots, namely $x = -5$, and $x = 2$ and to pass through $y = -2$ on the y-axis. Determine the unique polynomial.

[12] A quadratic polynomial passes through $y = 5$ on the y-axis, and $y = 1$ at $x = 1$. It also has a pair of complex roots, which are known. Can we determine the quadratic polynomial? State the reason.

[13] Consider a polynomial of order 4 given by
$$y = 2x^4 + 4x^3 - 14x^2 - 16x + 24$$

Rewrite this polynomial equivalently to a clustered form and factorized form.

[14] A quadratic polynomial, assuming $a>0$, is given
$$y = ax^2 + bx + c$$

(a) Derive the equation for the minimum value of y.
(b) Derive a criterion for this equation to have only complex roots.

[15] A quadratic polynomial, assuming $a<0$, is given
$$y = ax^2 + bx + c$$

(a) Derive the equation for the maximum value of y.
(b) Derive a criterion for this equation to have only complex roots.

[16] What conclusion can be drawn from the results of the two prior problems?

[17] For simplicity of discussions, consider four data points given by

$$x = [1.1, \quad 2.3, \quad 3.9, \quad 5.1]';$$
$$y = [3.887, 4.276, 4.651, 2.117]' ;$$

Determine the unique polynomial in power series form passing through the four points.

[18] Repeat [17] by first writing the polynomial in the Lagrange interpolation form, and then transforming to the power series. Hint: transform each term in the Lagrange formula to a power series using the **polyfit** command, and add them up together. The final answer must agree with that for [17].

---

[19] Revise Figure 2.6 for the 6th order interpolation polynomial. State what you observe.

[20] Revise Figure 2.7 using 7 Chebyshev points. State what you observe, particularly in comparison with the results of Problem [19].

Chapter 3
Numerical Integration Methods

Until computers were developed, mathematicians and scientists spent much time to figure out how to integrate mathematical functions. Today, however, with numerical integration methods, integration is very easy. The fundamental principle of numerical integrations is to fit an interpolation polynomial to the function to be integrated, and then integrate the polynomial.

Although there are numerous numerical integration methods, the trapezoidal rule and the Simpson's rule, described in sections 3.1 through 3.4, satisfy almost all numerical integration needs. The cases of involving singularity are explained in Section 3.5. In the final section, a method of double integration is described.

Octave/Matlab commands used in this section:
sum(a): If a is a column or row array, **sum** calculates the sum of all elements in the array, and if a is a 2-dimensional array, **sum** calculates the sum of each column of the array.
./ : array division operator

3.1 Trapezoidal Rule

The simplest numerical integration method is the trapezoidal rule. Suppose a function $y = f(x)$ is to be integrated from a to b as

$$I = \int_a^b f(x)dx \tag{3.1}$$

Defining the following notations,

$$f_1 = f(a) \quad \text{and} \quad f_2 = f(b)$$

the linear interpolation between the two end points is

60

$$g(x) = \frac{f_2 - f_1}{b - a}(x - a) + f_1 \qquad (3.2)$$

The area under the linear interpolation is

$$(b - a)\frac{f_1 + f_2}{2}$$

which is the trapezoidal rule approximation.

Although we may write the trapezoidal rule as

$$I \approx (b - a)\frac{f_1 + f_2}{2}$$

we prefer to include its error term as

$$I = (b - a)\frac{f_1 + f_2}{2} + E \qquad (3.3)$$

where E is the error term of the trapezoidal rule, and an equality sign is used rather than an approximation sign.

An expression for E is

$$E \approx -\frac{(b - a)^3}{12}\overline{f''} \qquad (3.4)$$

where $\overline{f''}$ is an average of the second derivative of f in the interval of integration. With the foregoing expression, we learn that the error of the trapezoidal rule is proportional to $(b - a)^3$, and also proportional to $\overline{f''}$. We observe here that the error is significantly affected by the size of the interval $b - a$. Indeed, as $b - a$ decreases, the error decreases very rapidly because the error is proportional to the cube of $b - a$.

To illustrate an application of the trapezoidal rule, we assume $f(x) = \cos(x)$, $a = 0$, and $b = 1.4$, so Eq.(3.1) becomes

$$I = \int_0^{1.4} \cos(x)dx \qquad (3.5)$$

In Figure 3.1, $\cos(x)$ and its linear interpolation are plotted, together with the results of the trapezoidal rule. In Figure 3.1, however, the following notations are used: $x_1 = a$, $x_2 = b$, $f_1 = f(a)$ and $f_2 = f(b)$. The exact I equals the area under the curve of $f = \cos(x)$, while the result of the trapezoidal rule is the area under the linear interpolation.

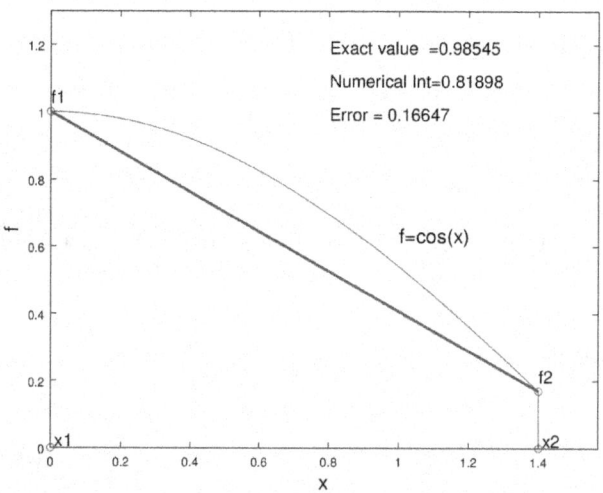

Figure 3.1 Trapezoidal rule with a single interval

The trapezoidal rule may be repeatedly applied by dividing the interval between a and b into multiple equispaced subintervals. The trapezoidal rule applied with multiple subintervals may be called in other literatures as extended trapezoidal rule. However, in this writing we do not use 'extended' for the sake of simplicity. With 2 subintervals, linear interpolation is applied in each subinterval as shown in Figure 3.2, and the trapezoidal rule becomes

$$I = \frac{h}{2}(f_1 + 2f_2 + f_3) + E \qquad (3.6)$$

where

$$h = \frac{b-a}{2}, \ x_i = a + h(i-1)$$

and

$$f_i = f(x_i)$$

The error E in Eq.(3.6) is

$$E \approx -\frac{(b-a)h^2}{12} \overline{f''} \qquad (3.7)$$

By comparing Eq.(3.4) and Eq.(3.7), the error with 2 subintervals become ¼ of the error with one interval, because $(b-a)h^2$ in Eq.(3.7) is ¼ of $(b-a)^3$ in Eq.(3.4): note that $\overline{f''}$ value in both equations are identical.

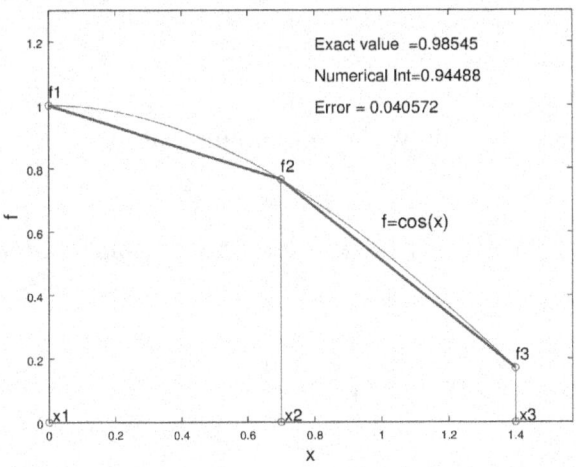

Figure 3.2 Trapezoidal rule with two subintervals

Likewise, with four subintervals, the trapezoidal rule becomes

$$I = \frac{h}{2}(f_1 + 2f_2 + 2f_3 + 2f_4 + f_5) + E \qquad (3.8)$$

where

$$h = \frac{b-a}{4}, \ x_i = a + h(i-1), \ f_i = f(x_i) \ \text{with} \ i = 1, 2..., 5$$

The results of the calculation for the same problem is shown in Figure 3.3.

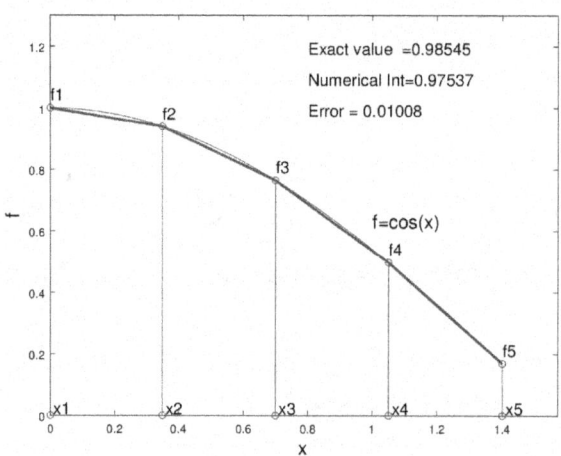

Figure 3.3 Trapezoidal rule with four subintervals

The trapezoidal rule may be written for any number of sub-intervals n as

$$I = \frac{h}{2}(f_1 + 2f_2 + 2f_3 + .. + 2f_n + f_{n+1}) + E \qquad (3.9)$$

where

$$h = \frac{b-a}{n}, \ x_i = a + h(i-1), \ f_i = f(x_i), \ i = 1, 2..., n, n+1$$

The error term E in the foregoing equation is given as

$$E \approx -\frac{(b-a)h^2}{12} f''$$

(3.10)

Notice here that error of the trapezoidal rule decreases to 1/4 as h is decreased to a half.

Example 3.1
In the present sample programming, we compute the integration of Eq.(3.5) using 4 subintervals, but n in the script can be changed to any other integer value.

```
%C3_1.m
n=4; a = 0; b = 1.4;
h=(b-a)/n;
x=a:h:b;
f=cos(x);
I = 0.5*h*(2*sum(f)-f(1)-f(length(f)))
```

ans = 0.97537

Example 3.2
We show now the results of the trapezoidal rule for an increased number of subintervals, n. Table 3.1 illustrates the results of the trapezoidal rule applied to Eq.(3.5).

The first column is n, the second is the results of the trapezoidal rule, the third their errors. Notice that error of the trapezoidal rule decreases approximately by a factor of 1/4 as n is doubled.

Table 3.1   Results of the trapezoidal rule with different numbers of subintervals, n.

n	I_n (Trapezoidal)	Error of I_n
1	8.189770e-001	1.664727e-001
2	9.448780e-001	4.057170e-002
4	9.753693e-001	1.008040e-002
8	9.829335e-001	2.516234e-003
16	9.848209e-001	6.288177e-004

32   9.852925e-001   1.571894e-004
64   9.854104e-001   3.929640e-005
128 9.854399e-001   9.824042e-006

3.2 Reduction of the error by Romberg method

The observation in the prior example leads us to the thought that, if four times of the error in I_n approximately equals that of $I_{n/2}$, a gross error of $(4I_n - I_{n/2})/3$ should become nearly zero. The fourth column in Table 3.2 is indeed $(4I_n - I_{n/2})/3$, and fifth is the errors of the fourth. It shows that $(4I_n - I_{n/2})/3$ becomes significantly more accurate than I_n. This method is called Romberg integration. A more analysis of this will be explained in the next section.

Table 3.2   Results of the trapezoidal rule and Romberg integration with different values of n: $R_n \equiv (4I_n - I_{n/2})/3$

n	I_n (Trapezoidal)	Error of I_n	R_n	Error of R_n
1	8.189770e-001	1.664727e-001		
2	9.448780e-001	4.057170e-002	9.868450e-001	-1.395311e-003
4	9.753693e-001	1.008040e-002	9.855331e-001	-8.336873e-005
8	9.829335e-001	2.516234e-003	9.854549e-001	-5.153470e-006
16	9.848209e-001	6.288177e-004	9.854501e-001	-3.212108e-007
32	9.852925e-001	1.571894e-004	9.854498e-001	-2.006195e-008
64	9.854104e-001	3.929640e-005	9.854497e-001	-1.253658e-009
128	9.854399e-001	9.824042e-006	9.854497e-001	-7.835055e-011

```
% C3_2.m    Trapezoidal rule and Romberg integration
clear, clf
I(1)=0;
n=1/2
for k=2:9
n=n*2;
a = 0; b=1.4;
x=a:(b-a)/n :b;
m=length(x);plot( [x(m),x(m)],[0,cos(x(m))])
Iexact=sin(b);
I(k) = (b-a)/n*(sum( cos(x)) - cos(0)/2 - cos(b)/2);
fprintf('%i  %e  %.3e  %e  %.3e\n',n,I(k), Iexact-I(k), ...
   (4*I(k)-I(k-1))/3, Iexact-(4*I(k)-I(k-1))/3 )
```

end

3.3 Simpson's 1/3 rule

Considering the integral of Eq.(3.1) again, we divide the interval of integration, $a \leq x \leq b$, into 2 subintervals, and consider three data points, $x_1 = a$, $x_2 = (a+b)/2$, and $x_3 = b$. For each of x values, we calculate f and denote $f_1 = f(x_1)$, $f_2 = f(x_2)$ and $f_3 = (x_3)$. A quadratic polynomial can be fitted to these three data points. By integrating the quadratic polynomial from $x_1 = a$ to $x_3 = b$, we get the Simpson's rule:

$$I = \frac{h}{3}(f_1 + 4f_2 + f_3) + E \tag{3.11}$$

with

$$h = (b-a)/2$$

where h equals the width of the subintervals. A graphic explanation of the Simson's rule is given in Figure 3.4. An approximate expression for the error of the Simpson's rule is

$$E \approx -\frac{(b-a)h^4}{180} f'''' \tag{3.12}$$

Here, Eq.(3.12) suggests a very interesting fact that if a polynomial of order 3 or less is integrated by the Simpson's rule, the result is exact because the error term of Eq.(3.12) vanishes.

```
% C3_3.m   Plots Simpson's 1/3 rule, 2 subintervals
% Figure 3.4
clear, clf
I(1)=0;
k=4
n=2;
figure(n), hold on
xlabel('x', 'fontsize', 16)
ylabel('f', 'fontsize', 16)
```

```
axis([0 pi/2 0 1.3])
a = 0; b=1.4;
x=a:(b-a)/n :b;
plot([-1 2],[0 0])
% plotting approximation
aa=polyfit(x,cos(x), length(x)-1)
xa=0:1.4/20:1.4
```

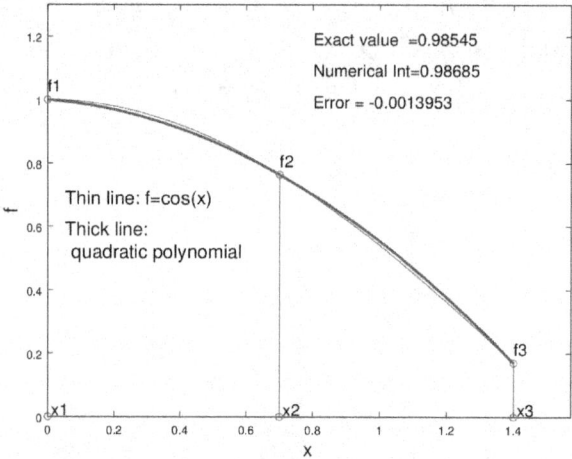

Figure 3.4   Graphic explanation of the Simpson's rule

(Continued)
```
ya=polyval(aa,xa)
plot( xa,ya, 'linewidth',2)
for m=1:length(x)-1
plot( [x(m),x(m)],[0,cos(x(m))])
end
m=length(x);plot( [x(m),x(m)],[0,cos(x(m))])
xp = a:0.01:b;
plot(xp,cos(xp))
text(0.05,0.7,'Thin line: f=cos(x)','fontsize',16)
text(0.05,0.6,'Thick line:','fontsize',16)
text(0.07,0.53,'quadratic polynomial','fontsize',16)
plot(x,cos(x),'o')
plot(x,0*(x),'o')
for m=1:length(x)
text(x(m),cos(x(m))+0.05,['f',num2str(m)],'fontsize',14)
text(x(m)+0.01,0.03,['x',num2str(m)],'fontsize',14)
end
```

```
Iexact=sin(b);
I(k) = (b-a)/n/3*( sum(cos(x)) + 3*cos(x(2)));
text(0.8, 1.2,['Exact value =', num2str(Iexact)], 'fontsize', 14);
text(0.8, 1.1,['Numerical Int=', num2str(I(k))], 'fontsize', 14);
text(0.8, 1.0,['Error =', num2str(Iexact-I(k))], 'fontsize', 14);
hold off
```

By dividing the whole range of integration into multiple couples of subintervals, and applying Eq.(3.11) to each couple of subintervals, the Simpson's rule becomes

$$I = \frac{h}{3}\left(f_1 + 4f_2 + 2f_3 + 4f_4 + 2f_5 + ... + f_{n+1}\right) + E \qquad (3.13)$$

where n is the total number of subintervals, and

$$h = (b-a)/n$$

An approximate formula for the error is the same as for the single application of the Simpson's rule:

$$E \approx -\frac{(b-a)h^4}{180} \overline{f''''} \qquad (3.14)$$

Equation (3.13) is called in other literature as Simpson's extended 1/3 rule, but in this writing, we do not use 'extended 1/3' for simplicity, but include '1/3' only when necessary to distinguish from the '3/8' rule which is explained in the next section.

Example 3.3
Programming and results
In the present sample programming shown below, we compute Eq.(3.13) using $n=2$ and 4 (subintervals).

Table 3.3 Results of Simpson's 1/3 rule with 2 and 4 subintervals:
n=2,   I=0.986845 Error=-0.001395
n=4,   I=0.985533 Error=-0.000083

69

```
%C3_4.m    Simpson's 1/3 rule
n=1; a = 0; b = 1.4;
for k=1:2
n=n*2;
h=(b-a)/n;
x=a:h:b;
f=cos(x);
fend=length(f);
I = h/3*( sum(f)+3*sum(f(2:2:fend-1))+ sum(f(3:2:fend-2)) );
fprintf('n=%f,  I=%f  Error=%f\n', n,I,0.98545-I)
end
```

Now we have the results of the Simpson's rule with $n = 2$ and $n = 4$ in Table 3.3, comparable to the results of the Romberg method of the trapezoidal rule in Table 3.2. Note that the error of the Simpson's rule with $n = 2$ is identical to the error of $(4I_n - I_{n/2})/3$ with $n=2$ in Table 3.2. Also the error of the Simpson's rule with $n = 4$ is identical to that of $(4I_n - I_{n/2})/3$ with $n = 4$ in Table 3.2. Why? The answer is that, if $(4I_n - I_{n/2})/3$ is expressed in terms of f_k's, it becomes identical with the Simpson's rule.

3.4 Simpson's 3/8 rule

The Simpson's rule described used three data points in its basic form: see Eq.(3.11). Then, what about integration of a cubic interpolation polynomial fitted to four data points? The integration scheme derived in this way, named the Simpson's 3/8 rule, is written as

$$I = \frac{3h}{8}(f_1 + 3f_2 + 3f_3 + f_4) + E \tag{3.15}$$

with

$$h = (b - a)/3$$

70

The error term is given by

$$E \approx \frac{3(b-a)h^4}{160} \overline{f''''}$$

(3.16)

Notice the foregoing error term has the same of order of h and $\overline{f''''}$ as in the error of the Simpson's 1/3 rule.

This scheme can be applied to multiple of 3 subintervals, but it is seldom used in this way. The most important application occurs when the total number of intervals is an odd number so the Simpson's 1/3 rule cannot be applied by itself. However, if the first three subintervals or the last three subintervals are integrated by the Simpson's 3/8 rule, then the rest can be integrated by the 1/3 rule. The 1/3 rule and 3/8 rule have the same order of accuracy, so they can be blended without sacrificing the order of accuracy.

3.5 Special cases of integration.

The trapezoidal rule and the Simpson's rule with multiple subintervals do great job in most times of integrating a function when the function has no singularity or if the interval of integration is finite. In this section, however, we pay attention to the special cases of integration between infinite limits, or integrating a function with singularity.

3.5.1 Integration with infinite limits
Consider the integration problem

$$I = \frac{1}{\sqrt{\pi}} \int_{-\infty}^{\infty} \exp(-x^2)dx$$

(3.17)

We replace the infinite limits, $-\infty$ and ∞, by finite limits, $-X$ and X. Then, the foregoing equation is rewritten

$$I \approx \frac{1}{\sqrt{\pi}} \int_{-X}^{X} \exp(-x^2)dx \qquad (3.18)$$

Here X is a value of x beyond which the contribution to the integral is negligibly small. If we take $X=10$, $exp(-10^2)$ becomes 3.7201e-44 so ignoring integration beyond $X=10$ should not cause any noticeable difference to the final results. Of course this assumption has to be examined and verified.

The numerical results of the trapezoidal rule with a total of 20 and 40 subintervals and $X=10$, are shown below:

n	I
20	1.00010344637241
40	1.00000000000000

This is awesome because the application of the trapezoidal rule is so simple, and the number of subintervals is not large. A question still remains, however, about the validity of the arbitrarily chosen value of X. For this problem, the exact analytical solution is easily available in a mathematical hand book which says the answer is unity, to which our numerical result agrees in double precision. So no further investigation is necessary for this problem. Generally, however, sensitivity analysis for the number of subintervals and the value of X is necessary.

Consider another example:

$$I = \int_{-\infty}^{\infty} \frac{1}{1+x^2} dx \qquad (3.19)$$

When plotted, this integrand looks similar to $\exp(-x^2)$ but approaches zero much more slowly than $\exp(-x^2)$, so we need to be more cautious.

We first set $X=10$, and do a sensitivity analysis for the effect of n. See the results in Table 3.4.

Table 3.4 Effect of n

n	X	h	I
40	20	0.50000	2.9472
80	20	0.25000	2.9447
160	20	0.12500	2.9435
320	20	6.2500e-002	2.9429
640	20	3.1250e-002	2.9426
1280	20	1.5625e-002	2.9424

```
%C3_5.m    Script for Table 3.4
n=40
for itr=1:6
X=10; h=2*X/n;
x = [-X:h:X];
I =  h * sum(1./ (1 + x.^2) );
fprintf('%i    %e    %e\n', n,h,I)
n=n*2;
end %for
```

Table 3.4 shows that the result is still changing even with the large number of subintervals, n=1280.

We next investigate the effect of changing X. Let us set h=20/1280 that is the smallest h used in Table 3.4. The results are in Table 3.5.

Table 3.5 Effect of X

X	h	I	Error of I
80	0.062500	3.116604e+000	0.024989
160	0.062500	3.129095e+000	0.012497
320	0.062500	3.135343e+000	0.006249
640	0.062500	3.138468e+000	0.003125
1280	0.062500	3.140030e+000	0.001562
2560	0.062500	3.140811e+000	0.000781
5120	0.062500	3.141202e+000	0.000391
10240	0.062500	3.141397e+000	0.000195
20480	0.062500	3.141495e+000	0.000098
40960	0.062500	3.141544e+000	0.000049
81920	0.062500	3.141568e+000	0.000024

```
%C3_6.m    Table 3.5
```

```
clear
X=40; n=1280; h=2*X/n;
for m=1:11
X=X*2;
x = [-X:h:X];
I =  h * sum(1./ (1 + x.^2) );
fprintf('%.0f %f %e %f\n',  X,h,I, pi-I)
end
```

Convergence is very slow and obviously this is not an easy problem. The reason for the difficulty is due to slow approach of the integrand to zero as x approaches $\pm\infty$. In other words, the contribution of the skirt range of the integrand does not diminish fast enough.

For such a problem a better approach is a coordinate transformation that decreases contribution of the skirt range. We choose here an exponential transformation given by

$$x = \sinh(z) \tag{3.20}$$

Differentiating $x = \sinh(z)$ yields

$$dx = \cosh(z)dz \tag{3.21}$$

Therefore, Eq.(3.19) becomes

$$I = \int_{-\infty}^{\infty} \frac{1}{1+(\sinh(z))^2} \cosh(z)dz \tag{3.22}$$

By replacing $\pm\infty$ by $\pm Z$, we calculate

$$I \approx \int_{-Z}^{Z} \frac{1}{1+(\sinh(z))^2} \cosh(z)dz \tag{3.23}$$

We do a sensitivity analysis for the effect of Z using a constant value of $h = 0.01$:

Z	h	I
10	0.010000	3.14159264538633
20	0.010000	3.14159265358978
40	0.010000	3.14159265358978

Foregoing results show that convergence with respect to Z is very fast, and the result converges at $Z=20$. Now, fixing Z at $Z=20$, we do a sensitivity analysis with respect to n:

Z	n	I
20	20	3.23261852950290
20	40	3.14224265513747
20	80	3.14159268085394
20	160	3.14159264633286

Obviously the exact solution is π, which is 3.14159265358979. The foregoing results indicate that the numerical solution with the exponential transformation is accurate to the 7th decimal place with $Z=20$ and $n=160$, although accuracy can be further improved. However, comparing the effort in the present work to the earlier attempt without the exponential transformation, the exponential transformation makes the integration significantly easier. Without the exponential transformation, it was not clear when the solution converges to the exact solution. With the exponential transfor-mation, since convergence is fast, we can easily see when the solution converges to the exact solution.

```
%C3_7.m    Exponential transformation
Z=10;h=0.01
format long
for k=1:5
z=-Z:h:Z; n=length(z)-1;
I =  h * sum(cosh(z)./ (1 + sinh(z).*sinh(z)) );
[Z,h,I,pi-I];
fprintf('%f, %f, %e \n', Z,h,I)
Z=Z*2;
end

Z=20;
format long
n=20;
for k=1:10
h=2*Z/n;
```

75

```
z=-Z:h:Z; n=length(z)-1;
I =  h * sum(cosh(z)./ (1 + sinh(z).*sinh(z)) );
%[Z,n,I,pi-I]
fprintf('%i, %i, %.14f \n', Z,n,I)
n=n*2;
end
```

3.5.2 Integration with singularity at one or both ends

In this section we consider integration of a function that has singularity at one or both integration limits. For example,

$$I = \int_0^1 \frac{1}{\sqrt{x}\,(\exp(x)+1)}\, dx \qquad (3.24)$$

$$I = \int_0^1 \frac{\exp(-x^2)}{\sqrt{1-x^2}}\, dx \qquad (3.25)$$

We use now a general form as

$$I = \int_a^b f(x)dx \qquad (3.26)$$

where we assume $f(x)$ is singular at $x = a$ or $x = b$, or both, yet the function is integrable in the interval. For such a problem we use coordinate transformation such that $[a, b]$ is mapped to $[-\infty, \infty]$ and then apply the trapezoidal rule as described in Subsection 3.5.1.

Consider a coordinate transformation given by

$$x = g(z) \qquad (3.27)$$

where g is a function of z, such that

$$a = g(-\infty), \ b = g(\infty) \qquad (3.28)$$

Equation (3.27) may be equivalently written as $x = x(z)$. Then the integration on the x-coordinate may be transformed to that on the z-coordinate as

$$I = \int_a^b f(x)\,dx = \int_{-\infty}^{\infty} f(x(z))\frac{dx}{dz}\,dz \qquad (3.29)$$

We choose the following exponential transformation,

$$x = [a + b + (b-a)\tanh(z)]/2 \qquad (3.30)$$

The derivative of Eq.(3.30) is

$$\frac{dx}{dz} = \frac{b-a}{2\cosh^2(z)} \qquad (3.31)$$

So, Eq. (3.29) becomes

$$I = \int_{-\infty}^{\infty}\left(f(x(z))\frac{b-a}{2\cosh^2(z)}\right)dz \qquad (3.32)$$

which we numerically integrate using the trapezoidal rule following the method described in Subsection 3.5.1.

Example 3.4
We now evaluate Eq.(3.25), or equivalently

$$I = \int_0^1 f(x)\,dx \qquad (a)$$

with

$$f(x) = \exp(-x^2)/\sqrt{1-x^2} \qquad (b)$$

After the exponential transformation of Eq. (3.30) with

$$a = 0,\ b = 1$$

the following relations may be verified easily:

$$z = -\infty \text{ for } x = 0, \text{ and } z = \infty \text{ for } x = 1 \qquad \text{(c)}$$

Then, the integration on the x-coordinate is transformed to that on the z-coordinate:

$$I = \int_a^b f(x)\,dx = \int_{-\infty}^{\infty} f(x(z))\frac{dx}{dz}\,dz \qquad \text{(d)}$$

On the z-coordinate, we replace $\pm\infty$ by $\pm Z$:

$$I \approx \int_{-Z}^{Z} f(x(z))\frac{dx}{dz}\,dz \qquad \text{(e)}$$

The trapezoidal rule applied to Eq.(e) is written as

$$I \approx h\sum_{i=1}^{n+1} f(x_i)\left(\frac{dx}{dz}\right)_i \qquad \text{(f)}$$

where we use the following relations:

$$h = 2Z/n \qquad \text{(g)}$$

$$z_i = -Z + (i-1)h \qquad \text{(h)}$$

$$x_i = [1 + \tanh(z_i)]/2 \qquad \text{(i)}$$

$$\left(\frac{dx}{dz}\right)_i = \frac{1}{2\cosh^2(z_i)} \qquad \text{(j)}$$

$$f(x_i) = \exp(-x_i^2)/\sqrt{1-x_i^2} \qquad \text{(k)}$$

In the actual computations we use $Z = 16$.

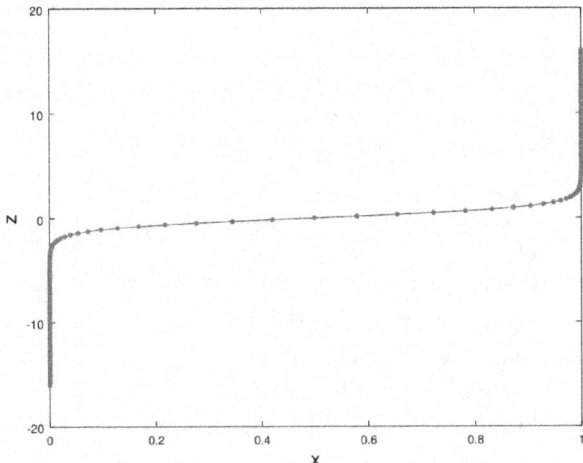

Figure 3.5 Relation between x and z points

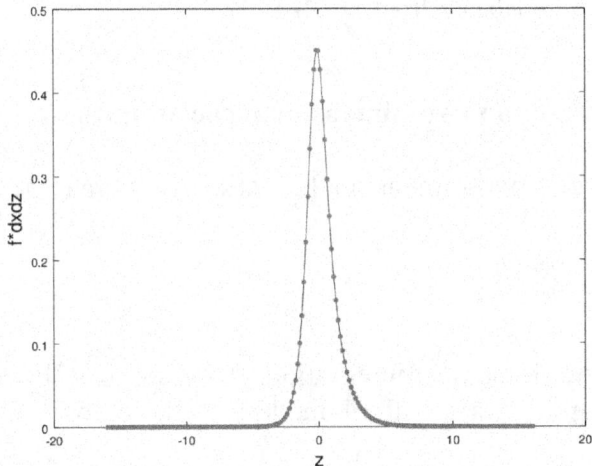

Figure 3.6 $f(x_i)\left(\dfrac{dx}{dz}\right)_i$ versus z_i

Calculations are performed by the script listed below. The final result is

$$I = 1.01321897949625$$

```
% C3_8.m    Computation of I
n=200;
```

```
Z=16;,
h=2*Z/n;
z=-Z:h:Z;
x=(1+tanh(z))/2;
dxdz=1./(2*cosh(z).*cosh(z));
f=exp(-x.^2)./sqrt(1-x.^2);
I=h*sum( f.*dxdz )
%
%plot fig 3.5
clf
figure(1)
plot(z,f.*dxdz, z,f.*dxdz,'.' )
xlabel('z','fontsize',16)
ylabel('f*dxdz','fontsize',16)
%
figure(2)
plot(x,z,x,z,'.')
xlabel('x','fontsize',16)
ylabel('z','fontsize',16)
```

3.6 Integration on two-dimensional coordinates

Integration of a two-dimensional function $f(x, y)$ may be written as

$$\iint_D f(x, y)dydx \tag{3.33}$$

where D is a domain of integration. Evaluation of Eq.(3.33) may not be straightforward if the boundary of the domain D is curved all the way.

If, however, it can be written in the form of

$$I = \int_a^b \int_{y=u(x)}^{y=v(x)} f(x, y)dxdy \tag{3.34}$$

where a and b are lower and upper bound on the x-axis; $y = u(x)$ and $y = v(x)$ are known functions which are respectively curves as lower and upper bounds in the y direction, then it can be evaluated by applying one of the numerical integration methods for one-dimensional problems.

To show the principle, we rewrite Eq.(3.34) to

$$I = \int_a^b F(x)dx \qquad (3.35)$$

where

$$F(x) = \int_{u(x)}^{v(x)} f(x,y)dy \qquad (3.36)$$

Equation (3.35) is a one-dimensional integration problem, if $F(x)$ can be evaluated for some number of points, $x = x_i$, on the x-axis.

Equation of (3.36) may be evaluated for selected values of $x = x_i$, as follows. Substituting $x = x_i$ into Eq.(3.36) yields

$$F(x_i) = \int_{u(i)}^{v(i)} f(x_i,y)dy \qquad (3.37)$$

where the following notations are used to avoid very small subscripts:

$$u(i) \equiv u(x_i), \quad v(i) \equiv v(x_i) \qquad (3.38)$$

Because $u(i) \equiv u(x_i)$, $v(i) \equiv v(x_i)$, and $x = x_i$ are all fixed values for each i, Eq.(3.37) is a one-dimensional integration problem, which can be evaluated using one of the numerical integration methods learned earlier.

Problems for Chapter 3

[1] Evaluate the following integrals using the trapezoidal rule with 2, 4, 8, 16, 32 and 64 subintervals:

(a) $\displaystyle\int_0^\pi \frac{dx}{2+\cos(x)}$

(b) $\displaystyle\int_1^2 \frac{\log_e(1+x)}{x}dx$

(c) $\displaystyle\int_0^{\pi/2} \frac{dx}{2+\sin^2(x)}$

[2] Apply the Romberg integration to the results of [1] and find the best answer for each integral.

[3] Evaluate the integrals in Problem [1] using the Simpson's rule with 2, 4, 8, 16, 32 and 64 subintervals:

[4] Evaluate the following integral analytically as well as by the Simpson's rule with 2 subintervals:

$$I = \int_0^2 (x^3 + 2x + 5)dx$$

What can you say about the comparison between the two results? And state why.

[5] Shown in the figure below is a cylinder of radius 0.5m and height 1m. The cylinder is cut at 45 degrees angle to the base. Calculate the total volume. Do not answer without numerical integration method even if you can find the answer from a geometrical observation.

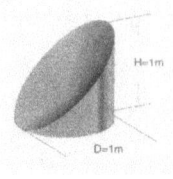

[6] The shape of a large parabolic arch is given by

$$y = 0.1x(30-x)$$

where y is the height and x is the measurement of the horizontal distance from one end, both in meters. Total curve length of the arch is given by

$$I = \int_0^{30} \sqrt{1 + \left(\frac{dy}{dx}\right)^2}\, dx$$

Evaluate I using 10 equally spaced subintervals in the x-direction.

[7] An automobile of mass M=5400kg is cruising at a speed of 30m/s. The engine is suddenly disengaged at t=0s. Assuming that the equation of coasting after t=0 is given by

$$M\frac{dv}{dt} = -8.276v^2 - 2000$$

where $v = v(t)$ is the speed (m/s) of the car at t. Using $dx/dt = v$, the left side can be written as $Mv(dv/dx)$. The first term on the right side is the aerodynamic drag, and the second term is the rolling resistance of the tires. Using the Simpson's rule with 2 subintervals, calculate how far the car travels until the speed reduces to 15m/s. Hint: The equation of motion is integrated as

$$I = \int_{15}^{30} \frac{5400}{8.276v^2 + 2000} v\, dv = \int_0^x dx' = x$$

[8] Evaluate the following integral accurately up to the sixth decimal place by the extended trapezoidal rule:

$$\int_{-\infty}^{\infty} \frac{\exp(-x^2)}{1+x^2}\, dx$$

[9] Evaluate the following integrals that involve singularity of the integrand accurately to the sixth decimal place by the trapezoidal rule using an exponential transformation:

$$\int_0^1 \frac{\tan(x)}{x^{0.7}}\,dx$$

$$\int_0^1 \frac{\exp(x)}{\sqrt{1-x^2}}\,dx$$

[10] Evaluate the following double integration problems:

(a) $$\int_1^2 \int_0^1 \sin(x+y)\,dy\,dx$$

(b) $$\int_0^1 \int_0^3 \sqrt{x+y}\,dy\,dx$$

[11] Evaluate the following double integration problem using the Simpson's rule with two subintervals for both x and y coordinates:

$$\int_2^7 \int_{y=\exp(-x)}^{y=\log(x)} \frac{\log_e(1+xy)}{x}\,dy\,dx$$

Chapter 4
Difference Approximations

Difference approximation refers to approximating derivatives of a function using the functional values at discrete points about where the derivative is to be evaluated. Difference approximations play central roles in solving ordinary differential equations and partial differential equations.

Octave/Matlab commands used in this chapter:
polyfit(x, y, n): fits a polynomial of order n to a data set, x and y
polyder(c): derivative of polynomial c

4.1 Difference approximations for first derivatives

We start with difference approximations for the first derivative of a function $f(x)$ at a selected point, $x = x_0$. The first derivative of a function is the gradient (or slope) of a function, so it can be approximately calculated if we have the values of the function at $x = x_0$ and at one more point $x_1 = x_0 + h$, where h is small increment of x. Then by drawing a linear interpolation passing through the two points, $(x_0, f(x_0))$ and $(x_0 + h, f(x_0 + h))$, we can find the slope of the linear interpolation, as illustrated in the first subplot of Figure 4.1.

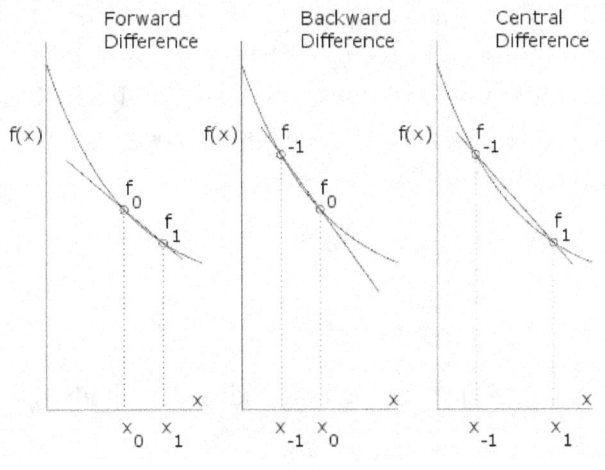

Figure 4.1 Graphical view of difference approximations (Subscrips -1, 0 and 1 refer to $x_0 - h$, x_0 and
$x_0 + h$, respectively)

The approximation for $f'(x_0)$ using the slope of the linear
interpolation is

$$f'(x_0) \approx \frac{f(x_0 + h) - f(x_0)}{h} \qquad (4.1)$$

called *forward difference approximation*. We can alternatively use
the point at $x_0 - h$ and calculate the slope as

$$f'(x_0) \approx \frac{f(x_0) - f(x_0 - h)}{h} \qquad (4.2)$$

which is called *backward difference approximation*. Furthermore,
using the two points, at $x_0 - h$ and $x_0 + h$, the slope

$$f'(x_0) \approx \frac{f(x_0 + h) - f(x_0 - h)}{2h} \qquad (4.3)$$

is called *central difference approximation*.

The graphic explanation of the difference approximations so
far is very intuitive and easy to understand, but lacks mathematical
rigor. The tool we will now use to analyze the difference
approximation is the Taylor expansion. The Taylor expansion of
$f(x_0 + h)$ about $x = x_0$ is

$$f(x_0 + h) = f(x_0) + hf'(x_0) + \frac{1}{2}h^2 f''(x_0)$$

$$+ \frac{1}{6}h^3 f'''(x_0) + \dots \qquad (4.4)$$

If we subtract $f(x_0)$ from the both sides and divide by h, Eq. (4.4)
becomes

$$\frac{f(x_0+h) - f(x_0)}{h} = f'(x_0) + \frac{1}{2}hf''(x_0) + \ldots \quad (4.5)$$

However, this is not an approximation, but an exact expression of what the forward difference formula is. In other words, the forward difference formula equals the first derivative $f'(x_0)$ plus additional terms.

We rewrite Eq. (4.5) one more time to

$$f'(x_0) = \frac{f(x_0+h) - f(x_0)}{h} - \frac{1}{2}hf''(x_0) + \ldots \quad (4.6)$$

or equivalently

$$f'(x_0) = \frac{f(x_0+h) - f(x_0)}{h} + E \quad (4.7)$$

with

$$E = -\frac{1}{2}hf''(x_0) + \ldots \quad (4.8)$$

Equation (4.7) is the forward difference approximation when E is ignored, which we do in actual use of the difference approximation. E represents the error of the forward difference approximation. E given by Eq. (4.8) is an infinite series, but the terms after the first term on the right side become increasingly smaller as h decreases. Therefore we ignore all terms except the first term, and write E as

$$E \approx -\frac{1}{2}hf''(x_0) \quad (4.9)$$

The backward difference approximation can be derived in a similar way, that is,

$$f'(x_0) = \frac{f(x_0) - f(x_0 - h)}{h} + E \tag{4.10}$$

with

$$E \approx \frac{1}{2} h f''(x_0) \tag{4.11}$$

We can see that the error of the backward difference approximation is the same as that of the forward difference approximation except the sign is opposite. A graphical interpretation of the backward difference approximation is shown in the second subplot of Figure 4.1.

We can guess that, by taking average of the forward difference approximation, Eq. (4.7), and the backward difference approximation, Eq. (4.10), the errors of each will be cancelled and the accuracy will increase. Indeed, that equals the central difference approximation. Of course the error of the central difference approximation is not zero, because the error terms for the forward difference and backward difference approximations are only approximation after truncating the higher order terms.

To find out the error term for the central difference approximation, we need to keep higher order terms in the Taylor expansions of $f(x_0 + h)$ and $f(x_0 - h)$, which are

$$f(x_0 + h) = f(x_0) + hf'(x_0) + \frac{1}{2} h^2 f''(x_0)$$

$$+ \frac{1}{6} h^3 f'''(x_0) + \dots \tag{4.12}$$

$$f(x_0 - h) = f(x_0) - hf'(x_0) + \frac{1}{2} h^2 f''(x_0)$$

$$- \frac{1}{6} h^3 f'''(x_0) + \dots \tag{4.13}$$

Subtracting the latter from the former and dividing by $2\,h$, we get

$$\frac{f(x_0 + h) - f(x_0 - h)}{2h} = f'(x_0)$$

$$+ \frac{1}{6} h^2 f'''(x_0) + \dots \qquad (4.14)$$

Rewriting the equation one more time, we get

$$f'(x_0) = \frac{f(x_0 + h) - f(x_0 - h)}{2h} + E \qquad (4.15)$$

with

$$E \approx -\frac{1}{6} h^2 f'''(x_0) \qquad (4.16)$$

Comparison of the above equation to those of the forward difference and backward difference approximations, we learn the error of the central difference approximation is one order higher than the forward or backward difference approximations. That is, while the error of the forward and backward difference approximations are proportional to h, the error of the central difference approximation is proportional to h^2. Thus, as h is decreased, the central difference approximation approaches the exact value faster than the forward difference or backward difference approximations.

Numerical accuracy of these three difference approximations is tested for a sample function $f(x) = \sin(x)$ at $x = \pi / 4$ and compared in Table 4.1.

Table 4.1 Comparison of backward difference, central difference and forward difference approximations

h	BW Diff	Error	Cent Diff	Error	FW Diff	Error
0.100000	0.741255	-0.034148	0.705929	0.001178	0.670603	0.036504
0.050000	0.724486	-0.017379	0.706812	0.000295	0.689138	0.017969
0.010000	0.710631	-0.003524	0.707095	0.000012	0.703559	0.003547
0.005000	0.708872	-0.001765	0.707104	0.000003	0.705336	0.001771
0.001000	0.707460	-0.000353	0.707107	0.000000	0.706753	0.000354

The foregoing results show that the central difference approximation is remarkably more accurate than other two approximations. The accuracy of the central difference approximation increases significantly faster than other two approximations as h becomes smaller.

```
%C4_1.m      Script for Table 4.1
x=pi/4;
exac=cos(x)
hh=[0.1, 0.05, 0.01, 0.005, 0.001];
for k=1:length(hh)
h=hh(k);
fd=(sin(x+h)-sin(x))/h;
cd=(sin(x+h)-sin(x-h))/2/h;
bd=(sin(x)-sin(x-h))/h;
if k==1
fprintf( 'h       BW Diff  Error   C Diff  Error   FW Diff   Error\n')
end %if
fprintf( '%f %f %f %f %f %f %f\n',h, bd, exac-bd, cd,exac-cd, fd, exac-fd)
end
```

In deriving a difference approximation, we can use more data points than 2. If we use 3 data points, for example, at x_0, $x_0 + h$ and $x_0 + 2h$, the difference approximation is derived as follows. We first expand $f(x_0)$, $f(x_0 + h)$ and $f(x_0 + 2h)$ into Taylor series:

$$f(x_0) = f(x_0) \tag{4.17}$$

$$f(x_0 + h) = f(x_0) + hf'(x_0) + \frac{1}{2}h^2 f''(x_0)$$
$$+ \frac{1}{6}h^3 f'''(x_0) + \dots \tag{4.18}$$

$$f(x_0 + 2h) = f(x_0) + 2hf'(x_0) + 2h^2 f''(x_0)$$
$$+ \frac{4}{3}h^3 f'''(x_0) + \dots \tag{4.19}$$

We multiply each equation by an undetermined constant a, b, c respectively and add up the three equations:

$$bf\left(x_0+h\right)+cf\left(x_0+2h\right) \; = \; (b+c)f\left(x_0\right) \; + \; (b+2c)hf'\left(x_0\right)$$

$$+ \; \frac{b+4c}{2}h^2 f''\left(x_0\right) \; + \; \frac{b+8c}{6}h^3 f'''\left(x_0\right) \; + \; ... \qquad (4.20)$$

where Eq.(4.17) times a did not contribute. In order to determine the 2 undetermined constants, namely b and c, we need 2 conditions. The first condition required is the coefficient of the second term on the right side is 1, namely

$$b + 2c \; = \; 1 \qquad (4.21)$$

The second is the third term on the right side to vanish, that is

$$b + 4c \; = \; 0 \qquad (4.22)$$

This is because, if the second derivative term is not zero, the error will involve the second derivative. On the other hand, if it is zero, the error will be from the third derivative term, and it is one order higher than the forward or backward difference approximations using two data points.

Solving these two equations yields

$$b = 2, \; c = -1/2 \qquad (4.23)$$

So, substituting these values into Eq.(4.20) yields

$$2f\left(x_0+h\right)-\frac{1}{2}f\left(x_0+2h\right)$$
$$= \; \frac{3}{2}f\left(x_0\right) \; + \; hf'\left(x_0\right) \; + \; 0 \; - \; \frac{1}{3}h^3 f'''\left(x_0\right) \; + \; ... \quad (4.24)$$

We rewrite the foregoing equation again to

$$f'(x_0) = \frac{-f(x_0 + 2h) + 4f(x_0 + h) - 3f(x_0)}{2h} + E \qquad (4.25)$$

$$E \approx \frac{1}{3}h^2 f'''(x_0)$$

The method using the Taylor expansion is very powerful in deriving difference approximations, which can be used for any order of derivatives and any combination of data points. See more applications in Appendix D.

Frequently used difference approximations are summarized in Appendix C.

4.2 Difference approximation for second derivatives

To derive a difference approximation for the second derivative, at least three data points are necessary. In the remainder of this section, we show how to derive the central difference approximation.

For the central difference approximation, we use $f(x_0 + h)$, $f(x_0)$ and $f(x_0 - h)$. The following equations are Taylor expansions of $f(x_0 + h)$ and $f(x_0 - h)$:

$$f(x_0 + h) = f(x_0) + hf'(x_0) + \frac{1}{2}h^2 f''(x_0)$$

$$+ \frac{1}{6}h^3 f'''(x_0) + \frac{1}{24}h^4 f''''(x_0) \ldots \qquad (4.26)$$

$$f(x_0 - h) = f(x_0) - hf'(x_0) + \frac{1}{2}h^2 f''(x_0)$$

$$- \frac{1}{6}h^3 f'''(x_0) + \frac{1}{24}h^4 f''''(x_0) \ldots \qquad (4.27)$$

By adding these two equation we can easily see that the second derivative terms on the right side remain but the first derivative and third derivative terms vanish. Rearranging the equation we get

$$f''(x_0) = \frac{f(x_0+h) - 2f(x_0) + f(x_0-h)}{h^2} + E \qquad (4.28)$$

with

$$E \approx -\frac{1}{12}h^2 f''''(x_0)$$

4.3 Other difference approximations

If difference approximations which are not derived in this chapter nor listed in Appendix C are looked for, there are easy ways of deriving them for any order of derivative, or even with non-uniformly spaced points on the x-axis. One is based on differentiation of Lagrange interpolation explained in Chapter 2, and another is by use of Taylor expansion. Both are described in detail in Appendix D.

4.4 Difference approximation for partial derivatives

Difference approximations for the two-dimensional or three-dimensional functions are essentially same as for ordinary derivatives. For example, the partial derivative of $f(x,y)$ with respect to x at (x_0, y_0) is the same as the ordinary derivative of $f(x, y_0)$, which is a one-dimensional function with the y value fixed to y_0. Therefore, all the difference approximations learned in the prior sections apply. For the remainder of this section, we illustrate the forward and central difference approximations for first partial derivatives, and some difference approximations for second partial derivatives.

Forward difference approximations for the first partial derivatives:

$$\frac{\partial}{\partial x} f(x,y)\Big|_{x=x_0, y=y_0} = \frac{f(x_0 + h, y_0) - f(x_0, y_0)}{h} + E \quad (4.29)$$

$$E \approx -\frac{1}{2} h \frac{\partial^2}{\partial x^2} f(x,y)\Big|_{x=x_0, y=y_0}$$

$$\frac{\partial}{\partial y} f(x,y)\Big|_{x=x_0, y=y_0} = \frac{f(x_0, y_0 + h) - f(x_0, y_0)}{h} + E \quad (4.30)$$

$$E \approx -\frac{1}{2} h \frac{\partial^2}{\partial y^2} f(x,y)\Big|_{x=x_0, y=y_0}$$

Central difference approximations:

$$\frac{\partial}{\partial x} f(x,y)\Big|_{x=x_0, y=y_0} = \frac{f(x_0 + h, y_0) - f(x_0 - h, y_0)}{2h} + E \quad (4.31)$$

$$E \approx -\frac{1}{6} h^2 \frac{\partial^3}{\partial x^3} f(x,y)\Big|_{x=x_0, y=y_0}$$

$$\frac{\partial}{\partial y} f(x,y)\Big|_{x=x_0, y=y_0} = \frac{f(x_0, y_0 + h) - f(x_0, y_0 - h)}{2h} + E \quad (4.32)$$

$$E \approx -\frac{1}{6} h^2 \frac{\partial^3}{\partial y^3} f(x,y)\Big|_{x=x_0, y=y_0}$$

Central difference approximations for the second partial derivatives:

$$\frac{\partial^2}{\partial x^2} f(x,y)\Big|_{x=x_0, y=y_0} = \frac{f(x_0 + h, y_0) - 2f(x_0, y_0) + f(x_0 - h, y_0)}{h^2} + E$$

$$(4.33)$$

$$E \approx -\frac{1}{12}h^2 \frac{\partial^4}{\partial x^4} f(x,y)\Big|_{x=x_0, y=y_0}$$

$$\frac{\partial^2}{\partial y^2} f(x,y)\Big|_{x=x_0, y=y_0} = \frac{f(x_0, y_0+h) - 2f(x_0, y_0) + f(x_0, y_0-h)}{h^2} + E$$

$$(4.34)$$

$$E \approx -\frac{1}{12}h^2 \frac{\partial^4}{\partial y^4} f(x,y)\Big|_{x=x_0, y=y_0}$$

Forward difference approximation for a cross derivative:

$$\frac{\partial^2}{\partial x \partial y} f(x,y)\Big|_{x=x_0, y=y_0}$$

$$= \frac{f(x_0+h_x, y_0+h_y) - f(x_0+h_x, y_0) - f(x_0, y_0+h_y) + f(x_0, y_0)}{h_x h_y}$$

$$+E \qquad (4.35)$$

$$E \approx -\frac{1}{2}h_x \frac{\partial^3}{\partial x^2 \partial y} f(x,y)\Big|_{x=x_0, y=y_0} - \frac{1}{2}h_y \frac{\partial^3}{\partial x \partial y^2} f(x,y)\Big|_{x=x_0, y=y_0}$$

Problems for Chapter 4

[1] Evaluate the first derivative of $y = \sin(x)$ at $x = 1$ by the forward, central and backward difference approximations using h=0.1, 0.05, 0.01 and 0.001. Compute the error of each answer by comparison to the exact value. Hint: make sure that error = exact value – approximation.

[2] Calculate df / dx, where $f = \sqrt{x}$, at x=1 by the forward difference, central difference and backward difference approximations with h=0.1, 0.01, 0.05 and 0.001. Evaluate the error of

each answer by comparison to the exact value. Hint: Make sure that error = exact value – approximation.

[3] Compute the error term of each approximation worked in [1] and compare to the error calculated using the exact derivative values.

[4] Determine the optimum value of α of the following difference approximation:

$$f_i' \approx \alpha(f_i - f_{i-1})/h + (1-\alpha)(f_i - f_{i-2})/2h$$

Hint: cancel the errors of the first and the second terms.

[5] Determine the optimum value of α of the following difference approximation:

$$f_i'' \approx \alpha(f_{i+1} + 2f_i - f_{i-1})/h^2 + (1-\alpha)(f_{i+2} + 2f_i - f_{i-2})/(2h)^2$$

Hint: cancel the errors of the first and the second terms.

[6] A function table is given by

x	f
-0.1	4.157
0	4.020
0.2	4.441

(a) Derive the best difference approximation to calculate $f'(0)$ with the data given in the table.
(b) What is the error term of the difference approximation derived for (a)?
(c) Calculate $f'(0)$ by the formula derived for (a).

[7] Assuming the abscissas of data points x_i are equispaced, derive the error term of the following difference approximation:

$$f_i' = \frac{f_{i+3} + 9f_{i+1} - 8f_i}{6h}$$

[8] Any difference approximation can be derived by differentiating a polynomial fitted to the given data. This is also true if the polynomial is written in the Lagrange interpolation form. We now assume that f_{-2}, f_{-1} and f are given. Derive the backward difference approximation for f_0' by differentiating the Lagrange interpolation. However, this choir is rather time consuming if all the work is done by hand calculations. So write a script to do the entire job using **polyfit** and **polyder** commands. Assume first the spacing between any consecutive two data points on the x-axis is 1. Divide the difference approximation derived by h.

[9] A function table is given below for $f(x, y)$:

y\x	0.0	0.5	1.0	1.5	2.0
0.0	0.0775	0.1573	0.2412	0.3309	0.4274
0.5	0.1528	0.3104	0.4767	0.6552	0.8478
1.0	0.2235	0.4547	0.7002	0.9653	1.2533
1.5	0.2866	0.5846	0.9040	1.2525	1.6348

(a) Evaluate $\partial f / \partial y$ at $x=1$ and $y=0$ using the forward difference approximation.

(b) Evaluate $\partial^2 f / \partial x^2$ at $x=1$ and $y=1$ using the central difference approximation.

(c) Evaluate $\partial^2 f / \partial x \partial y$ at $x=0$ and $y=0$ using the forward difference approximation.

Chapter 5
Finding Roots of Nonlinear Equations

Solutions of an equation, $f(x) = 0$, are called roots or zeros. One equation may have only one root or multiple roots. If the equation is a polynomial, Octave/Matlab can find its roots by the **roots** command, but otherwise one of different approaches is required. We will focus on three most useful methods in this chapter, which are graphic method, bisection method and Newton's iteration.

Octave/Matlab commands used in this chapter:
roots(c): computes roots of a polynomial c
axis: controls axis of a graph
contour: plots contour of two-dimensional function
feval(s,x): evaluates the function named s in string variable;
        x is the argument of the function

5.1 Graphic method

Whenever you seek a root of a function, it is recommended to plot the function, $y = f(x)$, first using the graphic tools of Octave/ Matlab. An intersection of the curve with the x axis is a root. By plotting a graph, you can see how many roots possibly exist. If multiple roots exist, you may be interested in only one of them, or all the roots. Graphic plotting gives you approximate value(s) of the root(s). Plotted graph also gives you very important information of what ranges of x other numerical methods should search to find the exact values of the roots, or an initial value for an iterative scheme.

Suppose you seek zeros of

$$\cos(x)\cosh(x) + 1 = 0 \tag{5.1}$$

in the range, $0 \le x \le 20$. To find the answers, we rewrite it to

---

$$y = \cos(x)\cosh(x) + 1 \qquad\qquad (5.2)$$

and plot in Figure 5.1 for $0 \le x \le 20$. The x values at which $y = 0$ are the solutions.

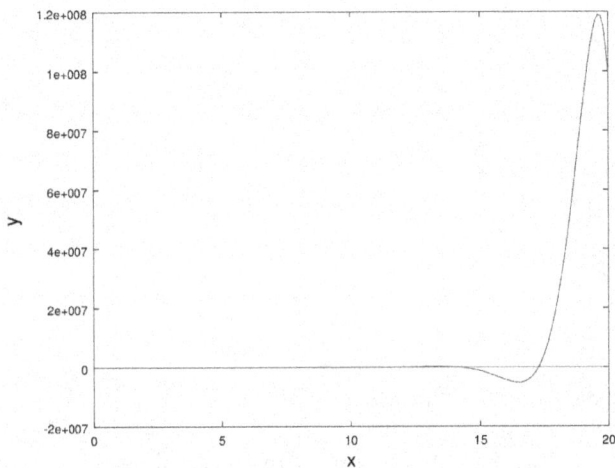

Figure 5.1 Plotting of $y = \cos(x)\cosh(x) + 1$

Figure 5.1 shows that the functional value becomes extremely large for $x > 15$, but one root is near $x = 17$. There may be other roots below $x = 15$, but they are hard to see. Therefore, we have to magnify the y-axis. For this purpose, we use the command, **axis([0, 20, -20, 20])** so the spaces beyond $y = \pm 20$ are eliminated.

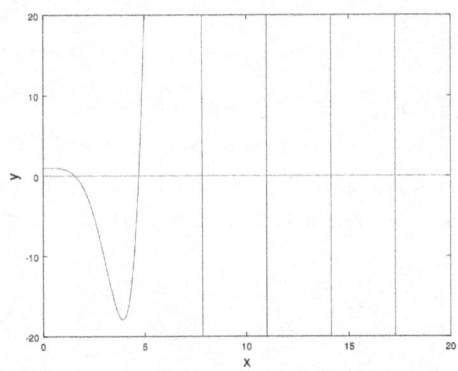

Figure 5.2 Revision of Figure 5.1 after using **axis** command

The new graph is in Figure 5.2, which shows there are 6 roots in the range.

It will help also if we divide the whole range, $0 \le x \le 20$, into smaller subintervals, and print out the subintervals that contain a root. With Octave/Matlab, this is done as follows:

```
%C5_1.m    Printing out subintervals
clear
xs=0:0.5:20; ys=cos(xs).*cosh(xs)+1;
n=0
for k=1:length(xs)-1
if  ys(k)*ys(k+1) <= 0;
n=n+1;
xlow(n)=xs(k);
xhigh(n)=xs(k+1);
end %if
end %for
fprintf('Subintervals containing a root %2.0f \n', n)
for nn=1:n
fprintf('xlow= %1.3f   xhigh=%1.3f \n', xlow(nn), xhigh(nn) )
end %for
```

The foregoing script found 6 subintervals that contain one root as shown in the following table:

Table 7.1 Subintervals containing one root

x-low	x-high
1.500	2.000
4.500	5.000
7.500	8.000
10.500	11.000
14.000	14.500
17.000	17.500

After finding small ranges of x in which a root is contained, the bisection method or Newton's iteration explained in the next sections may be used.

5.2 Bisection Method

―――

The bisection method works when an interval of x is known to have one root. We illustrate application of bisection method to find the root about $x = 5$ of the equation (which is different from the prior section):

$$f(x) = (1 - x\cos(x))x \tag{5.3}$$

A brief graphic analysis reveals that the root is contained between $a = 2$ and $c = 6$. The bisection method starts by calculating the midpoint b of the interval, which is $(a + c)/2 = 4$, namely $b = 4$. We examine which of the bisected subintervals $a \leq x \leq b$ or $b \leq x \leq c$ has the root. If $f(a)f(b) \leq 0$ the interval $a \leq x \leq b$ has it, and otherwise, if $f(b)f(c) \leq 0$, the interval $b \leq x \leq c$ has the root. Once the half interval that contains the root is found, the other is discarded. The lower bound of the new interval is now denoted by $x = a$, and the upper bound by $x = c$, and the same bisection procedure is repeated. The midpoint for the final iteration, b, is the answer for the root, with a possible maximum error of the final half width.

Computational results of the bisection method is shown in Table 5.1

Table 5.1   Results of the bisection method

It.	a	b	c	f(a)	f(c)	(c-a)/2
1	2.0000000,	4.0000000	6.0000000,	3.6645873,	-28.5661303	2.000e+000
2	4.0000000,	5.0000000	6.0000000,	14.4582979,	-28.5661303	1.000e+000
3	4.0000000,	4.5000000	5.0000000,	14.4582979,	-2.0915546	5.000e-001
4	4.5000000,	4.7500000	5.0000000,	8.7686149,	-2.0915546	2.500e-001
5	4.7500000,	4.8750000	5.0000000,	3.9016014,	-2.0915546	1.250e-001
6	4.8750000,	4.9375000	5.0000000,	1.0274563,	-2.0915546	6.250e-002
7	4.8750000,	4.9062500	4.9375000,	1.0274563,	-0.5042274	3.125e-002
8	4.9062500,	4.9218750	4.9375000,	0.2689399,	-0.5042274	1.562e-002
9	4.9062500,	4.9140625	4.9218750,	0.2689399,	-0.1158573	7.812e-003
10	4.9140625,	4.9179688	4.9218750,	0.0769936,	-0.1158573	3.906e-003
11	4.9140625,	4.9160156	4.9179688,	0.0769936,	-0.0193195	1.953e-003
12	4.9160156,	4.9169922	4.9179688,	0.0288652,	-0.0193195	9.766e-004
13	4.9169922,	4.9174805	4.9179688,	0.0047799,	-0.0193195	4.883e-004
14	4.9169922,	4.9172363	4.9174805,	0.0047799,	-0.0072680	2.441e-004
15	4.9169922,	4.9171143	4.9172363,	0.0047799,	-0.0012436	1.221e-004
16	4.9171143,	4.9171753	4.9172363,	0.0017682,	-0.0012436	6.104e-005

17	4.9171753,	4.9172058	4.9172363,	0.0002623,	-0.0012436	3.052e-005
18	4.9171753,	4.9171906	4.9172058,	0.0002623,	-0.0004906	1.526e-005
19	4.9171753,	4.9171829	4.9171906,	0.0002623,	-0.0001142	7.629e-006
20	4.9171829,	4.9171867	4.9171906,	0.0000741,	-0.0001142	3.815e-006
21	4.9171829,	4.9171848	4.9171867,	0.0000741,	-0.0000200	1.907e-006
22	4.9171848,	4.9171858	4.9171867,	0.0000270,	-0.0000200	9.537e-007
23	4.9171858,	4.9171863	4.9171867,	0.0000035,	-0.0000200	4.768e-007

Final result: Root $= 4.9171863$
|Maximum possible error| $= 4.768e\text{-}007$

The bisection scheme is graphically shown in Figure 5.3.

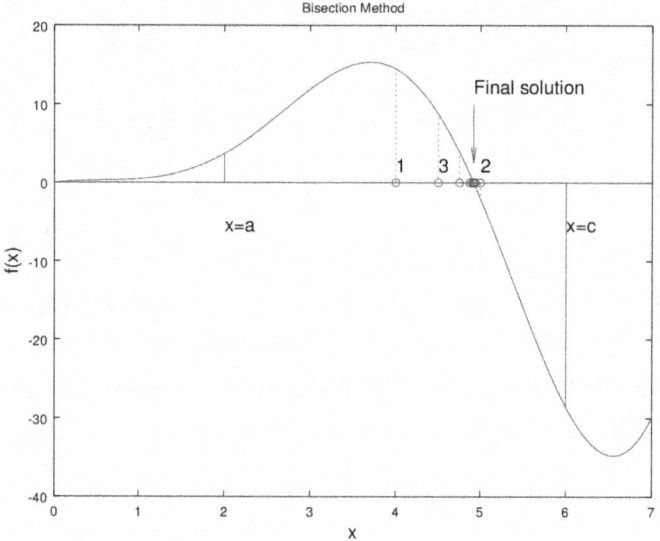

Figure 5.3 Graphic illustration of the bisection method

5.3 Newton's Iteration

We explain the Newton's iteration first on a graph shown in Figure 5.4. The Newton's iteration starts with an initial guess for the root x_0, which is marked by 0 in the graph. The value of $f(x_0)$ for this initial guess is calculated and marked by f_0. A tangential line is drawn from this point until the x-axis is reached at x_1. This point of x is denoted by 1 in the graph. The value of $f(x_1)$ for $x = x_1$ is calculated and denoted by f_1. A tangential line is drawn from this point until the x-axis is reached at x_2, where the x value is denoted by 2, and so on. The iteration is continued until the difference of the consecutive x values becomes within a criterion. The iterative results are shown in Table 5.2, which shows that the convergence is remarkably fast. Indeed, it becomes extremely fast toward the end of iterations.

Table 5.2   Results of Newton iteration
n=1.000000, x=6.080362, diff= 2.030362
n=2.000000, x=4.437722, diff=-1.642639
n=3.000000, x=5.066243, diff= 0.628521
n=4.000000, x=4.922307, diff=-0.143937
n=5.000000, x=4.917195, diff=-0.005111
n=6.000000, x=4.917186, diff=-0.000009
Final answer 4.917186

```
%C5_2.m     Newton iteration, Table 5.2
clear
x=4.05        %Initial guess setting
dx=0.001;
for n=1:6
f=(1-x*cos(x))*x;
xd=x+dx;
fd=((1-xd*cos(xd))*xd - f)/dx;
x=x-f/fd;
fprintf('n=%f, x=%f, diff=%f\n', n,x, -f/fd)
end %for
```

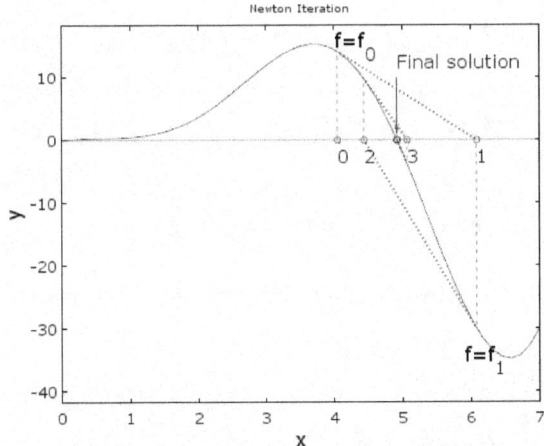

Figure 5.4   Newton iteration

```
%C5_3.m    Newton iteration, Fig 5.4
clf, clear
f_ans=newt_g('dem_bs',4.05 , 0, 7, 50);  %See function dem_g
text(6-0.1,-35,'f=f_1', 'fontsize',[18])
text(4+0.02, 15,'f=f_0','fontsize',[18])
set(gca, 'fontsize', 14)
xlabel('x' ,'fontsize',[18])
ylabel('y' ,'fontsize',[18])
fprintf('Final answer %e\n', f_ans)

% Function definition
function y = dem_bs(x)
y = (1 - x.*cos(x)).*x;

%Function newt_g
function x = newt_g(f_name, x0, xmin, xmax, n_points)
clf, hold off
% xmin, xmax:  min and max for plotting
del_x=0.001;
wid_x = xmax - xmin;  dx = (xmax- xmin)/n_points;
xp=xmin:dx:xmax;  yp=feval(f_name, xp);
plot(xp,yp); xlabel('x');ylabel('f(x)');
title('Newton Iteration'),hold on
f_name
ymin=min(yp); ymax=max(yp);wid_y = ymax-ymin;
yp=0.*xp;  plot(xp,yp)
x = x0;   xb=x+999; n=0;
while abs(x-xb)>0.000001
  if n>300 break; end
  y=feval(f_name, x);
    plot([x,x],[y,0],'--','linewidth',0.5);
    plot(x,0,'o')
  fprintf(' n=%3.0f, x=%12.6e, y=%12.6e\n', n,x,y);
  xsc=(x-xmin)/wid_x;
  if n<4,
    text(x, -wid_y/20, [ num2str(n)], 'fontsize', [16]),
  end
  y_driv=(feval(f_name, x+del_x) - y)/del_x;
  xb=x;
  x = xb - y/y_driv; n=n+1;
  plot([xb,x],[y,0],':','linewidth',2)
end
plot([x x],[0.02*wid_y 0.2*wid_y])
text( x, 0.25*wid_y, 'Final solution','fontsize', [16])
```

```
plot([x (x-wid_x*0.004)],[0.02*wid_y 0.05*wid_y])
plot([x (x+wid_x*0.004)],[0.02*wid_y 0.05*wid_y])
axis([xmin,xmax,ymin*1.2,ymax*1.2])
```

We now explain the theory of Newton's iteration. The initial value $x = x_0$ does not satisfy $f(x) = 0$ but we assume $x = x_0 + \delta x$ satisfies the equation, where δx is a correction. Expanding $f(x_0 + \delta x) = 0$ into a truncated Taylor series, we have

$$f(x_0 + \delta x) \approx f(x_0) + \delta x f'(x_0) \approx 0 \qquad (5.4)$$

where $f'(x_0)$ is the first derivative evaluated at $x = x_0$ and the approximation sign is used because the terms after the first order term are all truncated. The solution of Eq.(5.4) is

$$\delta x \approx -\frac{f(x_0)}{f'(x_0)} \qquad (5.5)$$

However, $x = x_0 + \delta x$ is not exact yet because we used a truncated Taylor expansion. Thus, we define $x_1 = x_0 + \delta x$ and we consider it as the next estimate for the repeated calculations. The sequence of iterative solutions is written as

$$x_m = x_{m-1} + \delta x = x_{m-1} - \frac{f(x_{m-1})}{f'(x_{m-1})} \qquad (5.6)$$

This iteration converges fast.

In using the Newton's iteration, the first derivative need not be derived analytically, because a numerical approximation for the first derivative may be used:

$$f'(x_{n-1}) \approx \frac{f(x_{n-1} + \Delta x) - f(x_{n-1})}{\Delta x} \qquad (5.7)$$

where Δx is a very small arbitrary number. In the script to create Table 5.2 listed after the table, Δx is set to 0.001.

A word of caution, however, is that Newton iteration may become unstable or diverge if the initial guess is too far from the solution. Such an unstable behavior can be easily experienced if Newton's iteration is graphically simulated with an initial guess of $x = 3$ in Figure 5.4, or the script after Table 5.2 is run with the same initial guess.

5.4 Newton's iteration for coupled implicit equations

Let us explain how Newton's iteration for coupled implicit equations works considering a set of simultaneous two-dimensional equations:

$$f(x, y) = 0 \qquad\qquad (5.8)$$
$$g(x, y) = 0 \qquad\qquad (5.9)$$

It is advised to plot the given equations in a graph so you can find how many roots are involved and read crude initial guess for the roots which Newton's iteration requires. Two-dimensional implicit functions cannot be plotted by **plot** command, but a method using **contour** explained in Appendix A works well.

We rewrite these equations to

$$f(x_0 + \delta x, y_0 + \delta y) = 0 \qquad\qquad (5.10)$$
$$g(x_0 + \delta x, y_0 + \delta y) = 0 \qquad\qquad (5.11)$$

where x_0 and y_0 are initial estimates and δx and δy are corrections to be determined. Expanding the foregoing equations into truncated first-order Taylor series yields

$$f(x_0 + \delta x, y_0 + \delta y) \approx f(x_0, y_0) + \delta x f_x + \delta y f_y \approx 0 \qquad (5.12)$$
$$g(x_0 + \delta x, y_0 + \delta y) \approx g(x_0, y_0) + \delta x g_x + \delta y g_y \approx 0 \qquad (5.13)$$

where f_x and f_y are partial derivatives of f with respect to x and y, respectively, and similarly g_x and g_y are partial derivatives of g with respect to x and y, respectively. The partial derivatives need not be derived analytically because we can approximate them as follows:

$$f_x \approx \frac{f(x_0 + \Delta x, y_0) - f(x_0, y_0)}{\Delta x} \qquad (5.14)$$

$$f_y \approx \frac{f(x_0, y_0 + \Delta y) - f(x_0, y_0)}{\Delta y} \qquad (5.15)$$

$$g_x \approx \frac{g(x_0 + \Delta x, y_0) - g(x_0, y_0)}{\Delta x} \qquad (5.16)$$

$$g_y \approx \frac{g(x_0, y_0 + \Delta y) - g(x_0, y_0)}{\Delta y} \qquad (5.17)$$

where Δx and Δy are small arbitrary numbers such as 0.001. Equations (5.12) and (5.13) are solved as a linear set of equations. Once we get the solutions for δx and δy, we will consider $x_1 = x_0 + \delta x$ and $y_1 = y_0 + \delta y$ as improved estimates. We repeat the same computation using x_1 and y_1 in place of x_0 and y_0. The iteration is repeated until convergence is reached.

Example
The two circles which are defined by the following equations are intersecting each other.

$$(x+4)^2 + (y-0.5)^2 = 36 \qquad \text{(a)}$$
$$(x-1)^2 + (y+1)^2 = 9 \qquad \text{(b)}$$

We will find the coordinates of the intersection above the x-axis. It is beneficial to plot the equations in a graph (see Figure 5.5) and read a crude approximation for the solution from the graph, which will be used as an initial guess for Newton's iteration.

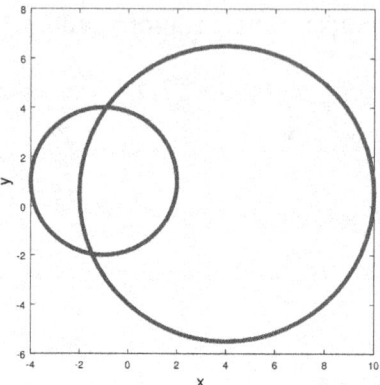

Figure 5.5 Plotting of the two equations

From the figure above, the root at a higher location is about $x = -1$ and $y = 4$, and the lower one is about $x = -1.5$ and $y = -2$. Since we are looking for the root above the x-axis, we pursue the former root further in the remainder of our effort.

To formulate the Newton iteration scheme, we define two functions as follows:

$$f(x, y) = (x+4)^2 + (y-0.5)^2 - 36 \qquad \text{(c)}$$
$$g(x, y) = (x-1)^2 + (y+1)^2 - 9 \qquad \text{(d)}$$

Using the first estimate for the root, $x_0 = -1$ and $y_0 = 4$, we write the following script:

```
%C5_4f1.m
function f=function_f(x,y)
f=(x-4)^2 + (y-0.5)^2 - 36;

%C5_4f2.m
function g=function_g(x,y)
```

108

```
g=(x+1)^2 + (y-1)^2 - 9;

%C5_4.m   Main part of the script
xest=-1;
yest=4;
dx=0.001; dy=0.001
%
for n=1:6
f= function_f(xest,yest);
g=function_g(xest,yest);
fx=( function_f(xest+dx,yest)-f)/dx;
fy=( function_f(xest,yest+dy)-f)/dy;
gx=( function_g(xest+dx,yest)-g)/dx;
gy=( function_g(xest,yest+dy)-g)/dy;
s=-[fx,fy; gx,gy]\[f,g]';
xest=xest+s(1); yest=yest+s(2);
fprintf('n=%f, x=%f, y=%f\n', n,xest,yest)
end
```

Result is as follows:

```
n=0, x=-1,          y=4
n=1, x=-0.875002, y=3.999979
n=2, x=-0.875259, y=3.997407
n=3, x=-0.875259, y=3.997406
n=4, x=-0.875259, y=3.997406
```

In the foregoing script, two functions are saved as function_f.m and function_g.m, respectively, in the directory where the main script is saved.

The script can work for any other set of coupled nonlinear equations if the definitions of the functions are altered. Only places to be changed are the two functions, and the values of **xest** and **yest**, which are initial estimates.

The Newton iteration explained in this section with the two coupled equations works for any number of implicit equations coupled, although finding appropriate initial guesses would become more difficult.

5.5 Successive substitution method

A nonlinear equation may be often written like a linear equation system

$$ax = y \tag{5.18}$$

where a is a matrix, and x and y are vectors, but the coefficients in matrix a are functions of the solution. Even y may also be dependent on x. For such a problem an iterative method may be written as

$$a^{(k-1)}x^{(k)} = y^{(k-1)} \tag{5.19}$$

where k is an iteration number. Before the iterative solution starts, $x^{(0)}$ has to be initialized by which $a^{(0)}$ and $y^{(0)}$ are calculated. Then the equation

$$a^{(0)}x^{(1)} = y^{(0)} \tag{5.20}$$

is solved as a linear equation. With $x^{(1)}$ thus obtained, $a^{(1)}$ and $y^{(1)}$ are calculated, and

$$a^{(1)}x^{(2)} = y^{(1)} \tag{5.21}$$

is solved. The iteration is continued until solution converges.

Problems for Chapter 5

[1] Develop a script that reads (1) a function to plot as input in a string form, (2) minimum and maximum values of x-axis for plotting and (3) number of points to be used for the plotting. Plot $\tan(x) - 2x + 0.2 = 0$, $0 \leq x \leq \pi/2 - 0.2$, to test the script.

[2] Determine approximately all positive roots of the following equations by the graphic method:

(a) $\tan(x) - 2x + 0.2 = 0$, $0 \le x \le pi/2 - 0.2$

(b) $\sin(x) - 0.3\exp(x) = 0$, $0 \le x \le 2$

(c) $\log_e(x) - 0.2x^2 + 1 = 0$, $0 \le x \le 5$

[3] Determine all roots of the following equations first by the graphic method, and then accurately by the bisection method:

(a) $0.5\exp(x/3) - \sin(x) = 0$, $0<x<10$

(b) $\log_e(1+x) - x^2/10 = 0$, $1 < x < 10$

[4] Evaluate $\tan^{-1}(3.5)$ in the interval of $0 \le x \le \pi$ using the bisection method. Hint: solve $\tan(x) = 3.5$.

[5] Determine all he roots of the equations in [2] accurately by Newton's iteration.

[6] Find the positive roots of the following equations by Newton's iteration:

(a) $0.5\exp(x/3) - \sin(x) = 0$, $x > 0$

(b) $\log_e(1+x) - x^2 = 0$, $x>0$

(c) $\exp(x) - 5x^2 = 0$

(d) $\sqrt{x+2} - x = 0$

[7] Find the solutions of the following coupled equations satisfying $x > 0$:

$$f_1(x,y) = x\exp(xy + 0.8) + \exp(y^2) - 3 = 0$$
$$f_2(x,y) = x^2 - y^2 - 0.5\exp(xy) = 0$$

Chapter 6
Curve Fitting To Measured Data

Curve fitting refers to fitting a function $g(x)$ to a set of data taken from experiment. The function $g(x)$ can be a polynomial, a linear combination of known functions or a nonlinear function, but it contains a certain number of undetermined constants. The number of data points will be denoted by L, which is in general a greater number than the number of undetermined constants k. The basic principle of determing the undetermined constants is to minimize the discrepancy between the data set and the fitting function, and named *least square method*.

Octave/Matlab commands used in this chapter:
polyfit(x, y, n): fits a polynomial of order n to a data set, x and y
polyder(c): derivative of polynomial c

6.1 Line fitting

Suppose we want to fit a line

$$g(x) = c_1 x + c_2 \qquad (6.1)$$

to a set of given data as illustrated in Table 6.1, where c_1 and c_2 are undetermined constants.

Table 6.1 A data set

x	y
0.1	0.61
0.4	0.92
0.5	0.99
0.7	1.52
0.7	1.47
0.9	2.03

Since the number of points $L=6$ (in Table 6.1) is greater than the number of unknowns (in Eq.(6.1)), we cannot satisfy for the fitted line to pass through every data point. We define residual by

$$r_i \equiv y_i - g(x_i) = y_i - (c_1 x_i + c_2) \tag{6.2}$$

which is the discrepancy between the data point and the fitted function at $x = x_i$. The sum of square of residuals is

$$R = \sum_i (r_i)^2 = \sum_i (y_i - c_1 x_i - c_2)^2 \tag{6.3}$$

The minimum of R occurs when partial derivatives of R with respect to each of c_1 and c_2 become zero:

$$\frac{\partial R}{\partial c_1} = -2 \sum_i x_i (y_i - c_1 x_i - c_2) = 0$$
$$\frac{\partial R}{\partial c_2} = -2 \sum_i (y_i - c_1 x_i - c_2) = 0 \tag{6.4}$$

Then Eq.(6.4) may be written in matrix form as

$$\begin{pmatrix} a_{1,1} & a_{1,2} \\ a_{2,1} & a_{2,2} \end{pmatrix} \begin{pmatrix} c_1 \\ c_2 \end{pmatrix} = \begin{pmatrix} z_1 \\ z_2 \end{pmatrix} \tag{6.5}$$

with

$$a_{1,1} = \sum_i x_i^2$$
$$a_{1,2} = a_{2,1} = \sum_i x_i$$
$$a_{2,2} = \sum_i 1 = L$$
$$z_1 = \sum_i x_i y_i$$

$$z_2 = \sum_i y_i$$

Equation (6.5) is a linear equation which we became familiar in Chapter 1.

The GNU Octave/Matlab solution for the data in Table 6.1 is as follows:

```
%C6_1.m
clf, clear
x=[0.1 0.4 0.5 0.7 0.7 0.9];
y=[0.61 0.92 0.99 1.52 1.47 2.03];
%c=polyfit(x, y, 1);
a(1,1)=sum(x.^2); a(1,2)=sum(x);  a(2,1)=a(1,2);
a(2,2)=length(x);
z(1)=sum(x.*y); z(2)=sum(y);
c=a\z';
g=polyval(c,x)
plot(x,y, '*', x, polyval(c,x))
axis([0 1 0 2.5])
xlabel('X', 'fontsize', 16)
ylabel('Y', 'fontsize', 16)
c'

c =
 1.76456   0.28616
```

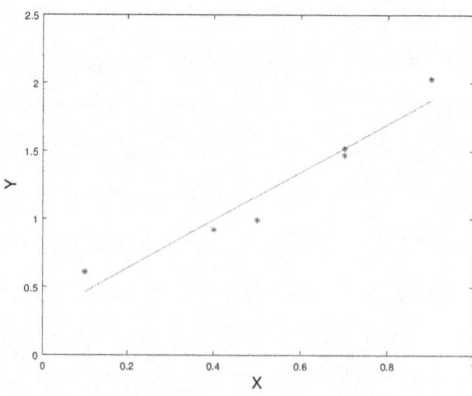

Figure 6.1   Linear regression with

the given data points

An equivalent way, which is slightly easier to formulate and solve, is to consider the problem as an over-determined linear equation. Consider Table 6.1 again. We write the following linear equations as if every data point is to be satisfied:

$$c_1 x_i + c_2 = y_i, \quad i = 1, 2, \ldots L$$

or equivalently in matrix form as

$$ax = y \quad (6.7)$$

with

$$a = \begin{pmatrix} 0.1 & 1 \\ 0.4 & 1 \\ 0.5 & 1 \\ 0.7 & 1 \\ 0.7 & 1 \\ 0.9 & 1 \end{pmatrix} \quad c = \begin{pmatrix} c_1 \\ c_2 \end{pmatrix} \quad y = \begin{pmatrix} 0.61 \\ 0.92 \\ 0.99 \\ 1.52 \\ 1.47 \\ 2.03 \end{pmatrix}$$

Equation (6.7) is an over-determined linear equation, which is un-solvable. To make this solvable we premultiply its transpose a' to Eq.(6.7) as

$$a'ac = a'y \quad (6.8)$$

Because $a'a$ is now a 2x2 matrix, and $a'y$ is 2 column vector, it is solvable as

$$c = (a'a) \backslash a'y \quad (6.8)$$

GNU Octave/Matlab script for this second way is

```
%C6_2.m
x=[0.1  0.4  0.5  0.7  0.7  0.9 ];
```

```
y=[0.61 0.92 0.99 1.52 1.47 2.03];
a = [x', zeros(6,1)+1]; y=y';
c=(a'*a)\(a'*y)
```

c =
    1.76456
    0.28616

Thus, the result of calculations for the coefficients of regression function is identical between the two ways of formulation.

However, GNU Octave/Matlab command, **polyfit(x, y, 1)**, finds the same results skipping all the details of the foregoing equations and labor:

```
%C6_3.m
x=[0.1 0.4 0.5 0.7 0.7 0.9];
y=[0.61 0.92 0.99 1.52 1.47 2.03];
c=polyfit(x, y, 1)
```

c =
    1.76456 0.28616

6.2 Nonlinear curve fitting with a power function

For some type of problems, the power function given by

$$g(x) = \beta x^{\alpha} \tag{6.7}$$

may fit the data better than other functions, where α and β are undetermined constants. To determine the constants we take logarithm of the equation:

$$\log(g) = \alpha \log(x) + \log(\beta) \tag{6.8}$$

So, if we redefine the terms in Eq.(6.8) as

$$G = \log(g)$$
$$c_1 = \alpha$$
$$X = \log(x)$$
$$c_2 = \log(\beta)$$

Eq.(6.8) becomes a linear equation,

$$G = c_1 X + c_2 \tag{6.9}$$

Then, the method written in the prior section applies to determine the two coefficients, c_1 and c_2. The constants α and β can be computed easily.

In applying the present method using Octave/Matlab, the data are written directly in the script below. The result of fitting is shown in Figure 6.2.

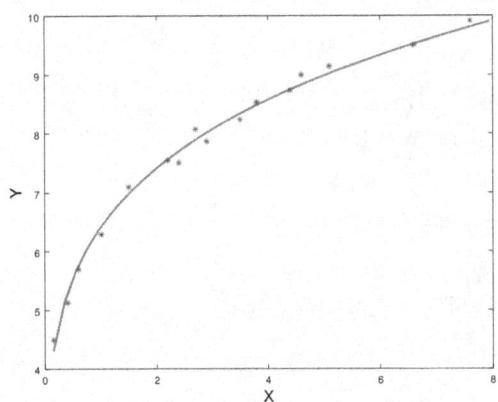

Figure 6.2 Curve fitting with a power function

```
%C6_4.m
clear, clf
x = [0.15,  0.4,   0.6,   1.01,  1.5,   2.2,   2.4,   ...
     2.7,   2.9,   3.5,   3.8,   4.4,   4.6,   5.1,   ...
     6.6,   7.6];
y = [4.4964,5.1284,5.6931,6.2884,7.0989,7.5507,7.5106, ...
     8.0756,7.8708,8.2403,8.5303,8.7394,8.9981,9.1450, ...
     9.5070,9.9115];
```

```
c = polyfit(log(x), log(y),1)
alpha=c(1)
beta=exp(c(2))
xp=0.15:0.2:8
plot(x, y, '*', xp, beta*xp.^alpha,'linewidth',2)
xlabel('X', 'fontsize', 16)
ylabel('Y', 'fontsize', 16)


c =
   0.20935  1.85876
alpha = 0.20935
beta = 6.4157
```

6.3 Curve fitting with a higher order polynomial

The method of least square written in Section 6.1 can be extended to a higher order polynomial fitting. An nth order polynomial is written as

$$g(x) = c_1 x^n + c_2 x^{n-1} + ... + c_{n+1} \qquad (6.10)$$

Using the notations defined in Section 6.1, residual of the curve for each data point can be written as

$$r_i = y_i - g(x_i), \quad i = 1, 2, ... L \qquad (6.11)$$

The sum of the square of residuals is

$$R = \sum_i r_i^2 \qquad (6.12)$$

In accordance with the principle of the least square method, we set the partial derivative of R with respect to each undetermined coefficient to zero:

$$\frac{\partial R}{\partial c_k} = 0, \quad k = 1, 2, ... n+1 \qquad (6.13)$$

which provides the number of linear equations equal to the number of unknowns, $n+1$. So a linear equation in the matrix form may be written in the form

$$ac = z \qquad (6.14)$$

where a is a square matrix, c is a column vector of c_1, c_2, .. c_{n+1}, and z is a column vector that come from y_i's. Solving Eq. (6.14) determines the coefficients c_i's.

Using GNU Octave/Matlab, the operation is simple:

>>c=polyfit(x, y, n)

where x and y are array forms of the data points, n is the order of the polynomial selected to fit the data.

In the following illustration, we use the same data as used in Section 6.1 and use a quadratic polynomial, namely $n=2$. A script is listed below, and the results are shown in Figure 6.3.

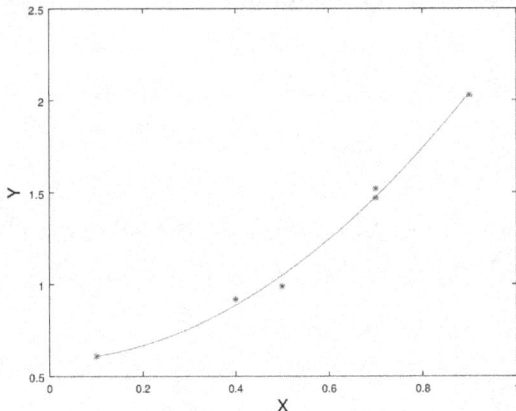

Figure 6.3   Curve fitting with a quadratic polynomial

```
%C6_7.m
clf, clear
x=[0.1 0.4 0.5 0.7 0.7 0.9];
y=[0.61 0.92 0.99 1.52 1.47 2.03];
```

```
c=polyfit(x, y, 2);
g=polyval(c,x);
xp=0.1:0.02:0.9;
plot(x,y, '*', xp, polyval(c,xp))
axis([0 1 0 2.2])
xlabel('X', 'fontsize', 16)
ylabel('Y', 'fontsize', 16)
```

6.4 Curve fitting with a linear combination of known functions

In fitting a function to data points, a linear combination of any known functions may be used:

$$g(x) = c_1 f_1(x) + c_2 f_2(x) + ... + c_n f_n(x) \qquad (6.15)$$

where f_1, f_2, ... are known functions, c_1, c_2, ... are undetermined constants, and n is the total number of known functions. The undetermined constants are determined by the least square method.

Example
Determine the coefficients of the function,

$$g(x) = c_1 + c_2 x + c_3 \sin(x) + c_4 \exp(x)$$

fitted to the data given in Table 6.1.

The solution is implemented in the script that follows. The result of fittng is shown in Figure 6.4:

```
%C6_8.m
clear; clf
data=[ 0.1  0.61;
       0.4  0.92;
       0.5  0.99;
       0.7  1.52;
       0.7  1.47;
       0.9  2.03];
x = data(:,1);   y = data(:,2);
A(:,1)=ones(size(x));   A(:,2)=x;   A(:,3)=sin(x);   A(:,4)=exp(x);
```

```
c = A\y;
xx = 0:0.01:1;
g= c(1)*ones(size(xx)) + c(2)*xx + c(3)*sin(xx) + c(4)*exp(xx);
axis('square');
plot(x, y,'*', xx, g,'linewidth', 2);
xlabel('x','fontsize',18); ylabel('y','fontsize',18)
axis([0, 1, 0, 3])
```

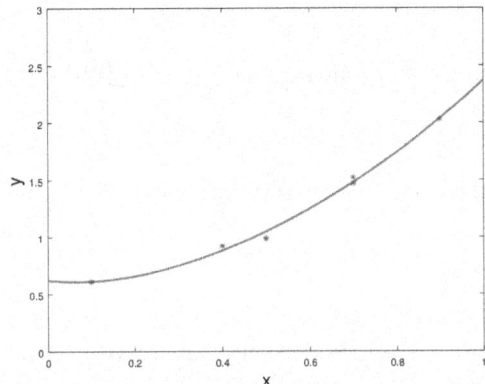

Figure 6.4 Curve fitting using linear combination of
known funcitons

Problems for Chapter 6

[1] Determine a line regression function fitted to the following
data:

x	y
1.0	2.0
1.5	3.2
2.0	4.1
2.5	4.9
3.0	5.9

[2] Write an over-determined equation for a linear equation for the
following data set:

x	y
0.1	9.9
0.2	9.2
0.3	8.4

0.4	6.6
0.5	5.9
0.6	5.0
0.7	4.1
0.8	3.1
0.9	1.9
1.0	1.1

Solve the over-determined linear equations. Verify the results with **polyfit**.

[3] The data set below is to be fitted by

$$y = \alpha \exp(\beta x)$$

Determine the constants. Plot the results with given data points.

x	y
0.0129	9.56;
0.0247	8.1845;
0.0530	5.2616;
0.1550	2.7917;
0.3010	2.2611;
0.4710	1.7340;
0.8020	1.2370;

[4] Fit polynomials of order 1, 2, and 3 to the following data set, and plot the determined polynomials:

x	y
0.002	0.618
0.004	1.1756
0.006	1.6180
0.009	1.9021

[5] Fit the function given by

$$g(x) = c_1 + c_2 x + c_3 \sin(\pi x) + c_4 \sin(2\pi x)$$

to the following data. Plot $g(x)$ with the data points.

x	y
0.1	0.0000
0.2	2.1220
0.3	3.0244
0.4	3.1399
0.6	2.8579
0.7	2.5140
0.8	2.1639
0.9	1.8358

Appendix A
Plotting of implicit functions

If a function is given implicitly, for example,

$$y^3 + \exp(y) = \tanh(x)$$

it cannot be expressed by x as a function of y, nor y as a function of x. Therefore, the methods to plot one-dimensional function, such as **plot**, do not work.

Plotting of an implicit function is possible, however, by the **contour** command. We rewrite the foregoing equation and define a two-dimensional function $z(x, y)$ as

$$z(x, y) = y^3 + \exp(y) - \tanh(x)$$

We plot its contour for only one level of $z = 0$. The plot of this function is illustrated in a figure below, plotted by the script attached.

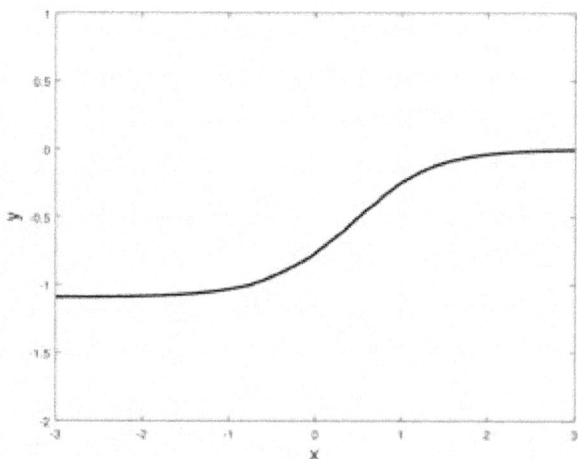

Figure A.1   Plot of $z = y^3 + \exp(y) - \tanh(x)$

```
%ApA_1.m      Plot of implicit function
clear, clf, cla
xm = -3:0.2:3;  ym = -2:0.2:1;  % input for meshgrid
[x, y] = meshgrid(xm, ym);
f = y.^3 + exp(y) - tanh(x);
%contour(x,y,f,[0,0],'linewidth',2) %Octave
 contour(x,y,f,[0,0])                %Matlab
xlabel('x', 'fontsize',16);
ylabel('y', 'fontsize',16)
```

The foregoing script can be adapted to other implicit functions easily. The mesh to cover the plotting domain will need to be changed by revising **xm** and **ym**. The definition of the implicit function in Line 5 also needs to be changed. Although you will not need any further instruction to plot your own implicit function, more information may be found in Reference [Nakamura 2015].

Appendix B
Inaccurate roots

Conversion of the power coefficients to the roots and vice versa is easy with the **roots** and **poly** commands, but we should be cautious for accuracy of the roots calculated by the **roots** command. This problem can occur when some of the roots are multiple or very close. To show an example, consider the polynomial,

$$y = (x-1)^6 = x^6 + 6x^5 + 15x^4 - 20x^3 + 15x^2 - 6x + 1 \qquad (2.32)$$

which has sextuple roots of $x = 1$. If we compute the roots from the power coefficients by the **roots** command, the answer is

```
>>r = roots([ 1 -6 15 -20 15 -6 1])
r =
   1.00296 + 0.00172i
   1.00296 - 0.00172i
   0.99998 + 0.00340i
   0.99998 - 0.00340i
   0.99706 + 0.00169i
   0.99706 - 0.00169i
```

Appendix C
List of difference approximations

The relation between the notations here and those used in Chapter 4 are:

$$x_i = x_0, f_i = f(x_0)$$
$$x_{i+1} = x_0 + h, f_{i+1} = f(x_0 + h)$$
$$x_{i+n} = x_0 + nh, f_{i+n} = f(x_0 + nh)$$

where the left side of each equation is the notation in this appendix, while the right side is used in Chapter 4.

First Derivative
Forward difference approximations

$$f_i' = \frac{f_{i+1} - f_i}{h} + E, \quad E \approx -\frac{1}{2} h f_i''$$

$$f_i' = \frac{-f_{i+2} + 4f_{i+1} - 3f_i}{2h} + E, \quad E \approx \frac{1}{3} h^2 f_i'''$$

$$f_i' = \frac{2f_{i+3} - 9f_{i+2} + 18f_{i+1} - 11f_i}{6h} + E, \quad E \approx -\frac{1}{4} h^3 f_i''''$$

Backward difference approximations

$$f_i' = \frac{f_i - f_{i-1}}{h} + E, \quad E \approx \frac{1}{2} h f_i''$$

$$f_i' = \frac{3f_i - 4f_{i-1} + f_{i-2}}{2h} + E, \quad E \approx \frac{1}{3} h^2 f_i'''$$

$$f_i' = \frac{11f_i - 18f_{i-1} + 9f_{i-2} - 2f_{i-3}}{6h} + E, \quad E \approx \frac{1}{4} h^3 f_i''''$$

Central difference approximations

$$f_i' = \frac{f_{i+1} - f_{i-1}}{2h} + E, \quad E \approx -\frac{1}{6}h^2 f_i'''$$

$$f_i' = \frac{-f_{i+2} + 8f_{i+1} - 8f_{i-1} + f_{i-2}}{12h} + E, \quad E \approx \frac{1}{30}h^4 f_i^{(5)}$$

($f^{(5)}$ is the 5th derivative)

Second Derivative
Forward difference approximations

$$f_i'' = \frac{f_{i+2} - 2f_{i+1} + f_i}{h^2} + E, \quad E \approx -hf_i'''$$

$$f_i'' = \frac{-f_{i+3} + 4f_{i+2} - 5f_{i+1} + 2f_i}{h^2} + E, \quad E \approx \frac{11}{12}h^2 f_i''''$$

Backward difference approximations

$$f_i'' = \frac{f_i - 2f_{i-1} + f_{i-2}}{h^2} + E, \quad E \approx hf_i'''$$

$$f_i'' = \frac{2f_i - 5f_{i-1} + 4f_{i-2} - f_{i-3}}{h^2} + E, \quad E \approx \frac{11}{12}h^2 f_i''''$$

Central difference approximations

$$f_i'' = \frac{f_{i+1} - 2f_i + f_{i-1}}{h^2} + E, \quad E \approx -\frac{1}{12}h^2 f_i''''$$

$$f_i'' = \frac{-f_{i+2} + 16f_{i+1} - 30f_i + 16f_{i-1} - f_{i-2}}{12h^2} + E,$$

$$E \approx \frac{1}{90}h^4 f_i^{(6)}$$

($f^{(6)}$ is the 6th derivative)

Third derivative
Forward difference approximation

$$f_i''' = \frac{f_{i+3} - 3f_{i+2} + 3f_{i+1} - f_i}{h^3} + E, \quad E \approx -\frac{3}{2}hf_i''''$$

Backward difference approximation

$$f_i''' = \frac{f_i - 3f_{i-1} + 3f_{i-2} - f_{i-3}}{h^3} + E, \quad E \approx \frac{3}{2}hf_i''''$$

Central difference approximation

$$f_i''' = \frac{f_{i+2} - 2f_{i+1} + 2f_{i-1} - f_{i-2}}{2h^3} + E, \quad E \approx -\frac{1}{4}h^2 f_i^{(5)}$$

($f^{(5)}$ is the 5th derivative)

Other difference approximations
Use one of the automatic methods for generating a difference approximation in Appendix D.

Appendix D
Automatic methods to find dfference approximations

Algorithm 1

The polynomial fitted to the k data points may be written in the Lagrange interpolation formula:

$$g(x) = \sum_{i=1}^{k} g_i(x) f_i$$

(D1)

where $g_i(x)$ is a polynomial passing through $(x_i, 1)$ and $(x_j, 0), j = 1,$ 2, .. k but $j \neq i$. The difference approximation for the nth derivative is obtained by differentiating Eq.(D1) n times and substituting $x = x_0$.

The whole operation becomes simpler if we set $x_0 = 0$, because the value of x_0 does not affect the result. We express the abscissa of a data point as a multiple of h, namely $x_i = a_i h$, where h is a unit spacing (or mesh spacing) on the x-coordinate, and a_i are assumed to be integers. (However, in applying the method developed now, a_i does not have to be an integer, but can be any fractional number.)

Using the coordinate transformation given by

$$x = hz$$

(D2)

the abscissas of the data points on the z-axis are $z = a_i$, and Eq.(D1) becomes

$$g(z) = \sum_{i=1}^{k} g_i(z) f_i$$

(D3)

where $g_i(z)$ is a polynomial such that $g_i(a_i) = 1$, but $g_i(a_j) = 0$ for $j \neq i$. On the z-axis, the difference approximation is obtained by differentiating $g(z)$ and evaluating at $z = 0$. (Readers are suggested to plot $g_i(z)$ on several sets of the data points, a_i.)

The derivatives on the x-axis and those on the z-axis are related by

$$dx = hdz$$

$$\frac{dg(x)}{dx} = \frac{1}{h}\frac{dg(z)}{dz}$$

$$\frac{d^2g(x)}{dx^2} = \frac{1}{h^2}\frac{d^2g(z)}{dz^2}$$

...

$$\frac{d^ng(x)}{dx^n} = \frac{1}{h^n}\frac{d^ng(z)}{dz^n}$$

The coefficients of the polynomial in the power series form of $g_i(z)$ is determined by **polyfit** for each i. The polynomial $g_i(z)$ is differentiated n times by **polyder** command. Then, the nth derivative of $g_i(z)$ is evaluated for $z = 0$, and divided by h^n, that becomes the coefficient of f_i in the difference approximation derived.

The following script computes the equations thus described. The user has to change the first three lines after the comment lines for each of different difference approximations: DerivOrder is the order of the derivative for which the difference approximation is to be found, m is a multiplier (input) that adjusts the numbers in both numerator and denominator in the difference approximation so only integers appear in the final formula, and $z=[$ ...$]$ are the data points to be used in the difference approximations. The values in

[...] are abscissas on the z-coordinate, or equivalently the abscissas on the x-coordinate divided by h.

```
%ApD_1.m     Automatic Difference Approximation Finder
%User input
clear
DerivOrder=1;
m=12;              %Multiplier in the final format
a=[ -2 -1 0 1 2];
fprintf(' Order of derivative=%i; Multiplier=%i\n', DerivOrder, m)
 fprintf(' Number of points=%i \n',length(a))
fprintf(' Points:\n')
a
% End of user input
for i=1:length(a)
y=a*0;
y(i)=1;
c=polyfit(a,y,length(a)-1);
for L=1:DerivOrder
c=polyder(c);
end %for
d(i)=polyval(c,0);
fprintf('Coefficient for a(%i) : %f/%ih^%i\n', a(i),
d(i)*m,m,DerivOrder)
end %for
```

Order of derivative=1; Multiplier=12
 Number of points=5
 Points:
 z =
  -2 -1 0 1 2
 Coefficient for f(-2) : 1.000000/12h^1
 Coefficient for f(-1) : -8.000000/12h^1
 Coefficient for f(0) : -0.000000/12h^1
 Coefficient for f(1) : 8.000000/12h^1
 Coefficient for f(2) : -1.000000/12h^1

The foregoing results are interpreted as

$$f_0' \approx \frac{f_{-2} - 8f_{-1} + 8f_1 - f_2}{12h}$$

This algorithm works even when the data points are not equispaced, and even when the abscissas are not integer multiple of h.

Algorithm 2

Suppose L data points are used and they are numbered like $i = \alpha, \beta, .., \lambda$. We assume $L \geq p + 1$, where p is the order of the derivative. The abscissas of the data points are $x_i = \alpha h, \beta h, .., \lambda h$ where h is a given constant. If $i = \alpha, \beta, .., \lambda$ are consecutive integers, h becomes the interval between two consecutive points. However, in the difference approximations we derive here, $i = \alpha, \beta, .., \lambda$ do not have to be integers nor do not have to increase at a constant increment.

The difference approximation for the pth derivative of $f(x)$ at $x = 0$ may be written as

$$f_0^{(p)} = \frac{a_\alpha f_\alpha + a_\beta f_\beta + ... + a_\lambda f_\lambda}{h^p} + E \tag{D4}$$

where a_i are L undetermined coefficients, $f_i = f(x_i)$, $i = \alpha, \beta$, .., λ are ordinates of the data points, $f_0^{(p)}$ is the pth derivative of $f(x)$ at $x = 0$, and E is the error term given by

$$E \approx c_1 h^{L-p} f_0^{(L)} + c_2 h^{L-p+1} f_0^{(L+1)} \tag{D5}$$

In the equation for E, the second term is ignored if $c_1 \neq 0$, but if $c_1 = 0$, the second term becomes the error term.

The essence of the algorithm is that the f values in Eq.(D4) are expanded into Taylor series about $x=0$, and the undetermined coefficients are determined so the error term becomes in as high power of h as possible.

For simplicity of explanation, we assume $p = 1$, $L = 3$ and $x_i = 0, h, 2h$:

$$f_0^{(1)} = \frac{a_0 f_0 + a_1 f_1 + a_2 f_2}{h} + E \tag{D6}$$

Substituting the Taylor expansions

$$f_1 = f_0 + h f_0' + \frac{1}{2} h^2 f_0'' + \frac{1}{6} h^3 f_0''' + \frac{1}{24} h^4 f_0'''' \cdots$$

$$f_2 = f_0 + 2h f_0' + \frac{4}{2} h^2 f_0'' + \frac{8}{6} h^3 f_0''' + \frac{16}{24} h^4 f_0'''' \cdots$$

into Eq.(D6) yields

$$f_0^{(1)} = \frac{1}{h}(a_0 + a_1 + a_2) f_0$$
$$+ (0 + a_1 + 2a_2) f_0'$$
$$+ \frac{h}{2}(0 + a_1 + 4a_2) f_0''$$
$$+ \frac{h^2}{6}(0 + a_1 + 8a_2) f_0'''$$
$$+ \frac{h^3}{24}(0 + a_1 + 16a_2) f_0''''$$
$$+ \ldots + E \tag{D7}$$

The foregoing equation has three undetermined coefficients, a_0, a_1 and a_2, so three conditions have to be imposed into it. We impose that the first term on the right side be zero, the second term equals the left side, and third term be zero, namely:

$$a_0 + a_1 + a_2 = 0$$
$$a_1 + 2a_2 = 1 \tag{D8}$$
$$a_1 + 4a_2 = 0$$

Its solution is $a_0 = -3/2$, $a_1 = 2$, and $a_2 = -1/2$. Substituting the solution back to Eq.(D7), we get

$$f_0^{(1)} = 0 + f_0' + 0 + \frac{h^2}{6}(-2)f_0''' + \frac{h^3}{24}(-6)f_0'''' + ... + E \qquad \text{(D9)}$$

Because $f_0^{(1)} = f_0'$, the foregoing equation becomes

$$E = \frac{h^2}{3}f_0''' + \frac{h^3}{4}f_0'''' + ...$$

or

$$E \approx \frac{h^2}{3}f_0''' \qquad \text{(D10)}$$

Also Eq.(D6) becomes

$$f_0' = \frac{-3f_1 + 4f_2 - f_3}{2h} + E \qquad \text{(D11)}$$

A script for the present algorithm is listed next.

```
% ApD_2.m     Difference approximation finder, algorithm 2
while 1
clear, clf
  fprintf( '\n==================================\n' );
  fprintf( ' Difference Approximation Finder  \n' );
  while 1
    km = input( '** Number of points ? ' );
    if km>1 break; end
    fprintf(' Input is invalid: Repeat.\n')
  end
%
  while 1
    fprintf( 'Input the point indices in row vector form ')
    fprintf( 'like [x x ... x]'); el = input('');
    if length(el) == km;  break; end
    fprintf( ' Number of points do not match with indices')
    fprintf( ' Repeat your input for indices.')
  end
```

```
       kdr = input('** Order of difference scheme to be derived ?  ' );
       z = 1.0; for  i = 1:kdr;  z = z*i; end
       for k = 1:km+2;   a(k,:) = el.^(k-1); end
       M = a(1:km, 1:km);
       rs = zeros(1,km)';  rs(kdr+1) = z;
%  kmp2 = km + 2;
     y = M^(-1)*rs;
              c = a*y;
       u = abs(y);
       for k = 1:km+2
          if k<=km; if u(k)<0.000001, u(k) = 1000;end; end
          if( abs( c(k) ) < 0.00000001 ) c(k) = 0;end
       end
       f_min = min(u);
              cf = y/f_min;
       fprintf( '\nDifference scheme:\n' );
       for  k = 1:km
          finv = 1.0/f_min;
          fprintf( ' +(%8.5f/( %8.5f h^%1.0f))*', cf(k),finv, kdr)
          fprintf( 'f( %3.1fh ) \n', el(k) );
       end
       fprintf('\nError term\n');
       dd = 1.0;
       for k = 1:km
           dd = dd*k;
       end
       for k = km+1:km+2,
         cm = -c(k);
         %km1 = k - 1;
         nh = k-1-kdr;
         if( k == km+1 & cm ~= 0 )
             fprintf( '   (%7.3f/%7.3f)h^%1.0f f', cm, dd, nh );
             for (i=1:k-1)   fprintf( '' );
             end
             break
         end
         if( k == km + 2 ),
           fprintf( '\n  +(%7.3f/%7.3f)h^%1.0f f', cm, dd, nh );
           for i=1:k-1
             fprintf( '' );
           end
         end
         dd = dd*k ;
       end
       fprintf('\n=====================================')
       kont = input( 'Type 1 to continue, or 0 to stop:' );
       if kont ==0, break; end
```

end

Difference Approximation Finder
** Number of points ? 5
Input the point indices in row vector form like [x x ... x]   [-2 -1 0 1 2]
** Order of difference scheme to be derived ?   1

Difference scheme:
 +(1.00000/(12.00000 h^1))*f(-2.0h)
 +(-8.00000/(12.00000 h^1))*f(-1.0h)
 +(0.00000/(12.00000 h^1))*f(0.0h)
 +(8.00000/(12.00000 h^1))*f(1.0h)
 +(-1.00000/(12.00000 h^1))*f(2.0h)

Error term
   (4.000/120.000)h^4 f`````

The script was run for the same contents of input as used in the script of Algorithm 1. The results are interpreted as follows:

$$f'(0) = \frac{f(-2h) - 8f(-h) + 8f(h) - f(2h)}{12h} + E \tag{D15}$$

or equivalently

$$f_0' = \frac{f_{-2} - 8f_{-1} + 8f_1 - f_2}{12h} + E$$

where

$$E \approx \frac{1}{30} h^4 f'''''\tag{D16}$$

One advantage of this script is that the error term is derived also. It is a very smart program. Indeed, all the difference approximations in this book were tested with this program before printing.

Both algorithms work for non-equispaced data points, and even when the data points are not integer multiple of unit spacing h.

Norm and condition number

The condition number of a matrix a can be computed in Octave/ Matlab simply by the **cond(a)** command. However, if you ask how the condition number is calculated, the answer is not so simple because the condition number depends on norm of the matrix, and there are several choices in the definition of the norm.

Indeed, **cond(a)** is a short version of the full command **cond(a, p)** where p can be chosen from '1', '2', 'inf' and 'fro'. In Octave/ Matlab, **norm(a)** returns the 2-norm of matrix a by default (p=2).

The 2-norm is also called spectral norm, and defined by

$$\|a\| = \sqrt{\lambda_{max}(a'a)} \tag{E1}$$

where $\lambda_{max}(a'a)$ is the largest eigenvalue of $a'a$, and where a' is transpose of a.

Example
For illustration purpose, let us define

$$a = \begin{pmatrix} 2 & -1 \\ -2 & 5 \end{pmatrix}$$

In Octave/Matlab, $\lambda_{max}(a'a)$ is computed by
```
>> a=[2,-1;-2 5]; lambda=max(eig(a'*a))
   lambda = 32
```

The 2-norm for a is
$$\|a\| = \sqrt{\lambda_{max}(a'a)} = \sqrt{32} = 5.6569$$

The condition number is defined by

$$\text{cond}(a) = \|a\| \|a^{-1}\| \qquad\qquad (E2)$$

In order to calculate the condition number without using the **cond** command, we need $\|a^{-1}\|$, which is computed in the following example.

Example
Here we compute $\lambda_{max}((a^{-1})'a^{-1})$:
```
>> a=[2,-1;-2 5]; lambda_inv=max(eig(inv(a)'*inv(a)))
   lambda_inv =  0.50000
```
$$\|a^{-1}\| = \sqrt{\lambda_{max}((a^{-1})'a^{-1})} = \sqrt{0.5} = 0.70711$$

Therefore, the condition number is $0.56569*0.70711=4.0000$

Now we use **cond(a)** to verify the foregoing calculations:
```
>>cond(a)
ans =  4.0000
```
(Excellent agreement.)

Answer of Problems

Problems for Chapter 1

[1]
```
>> a*b
ans =
   16
    9
    1
```

[2]
```
>> a*b
ans =
   14   5    1
   14   7   -5
    9   7   -1
```

[3]

(a)
```
>>[1 -1; 3  2]\[5; 7]
ans=
   3.4   -1.6
```

(b)
```
>>[ 3 1 2; -1 1 4; 3 1 2]\[2; 0; 5]
ans =
   0.97222   0.19444   0.19444
```

[4]

$$a^{-1} = \begin{pmatrix} 1/4 & 0 \\ 0 & 1/5 \end{pmatrix}, \quad b^{-1} = \begin{pmatrix} 1/3 & 0 & 0 \\ 0 & 1 & 0 \\ 0 & 0 & 1/2 \end{pmatrix}$$

[5]
```
>>cond([3.2406   2.9155; -9.7128  -8.7464])
ans = 7325.3
```

Octave solution
```
2028.28165815452    -2252.04923389317
```

Single precision Fortran
```
2028.084   -2251.830
```

Octave/Matlab answer is good, but single precision Fortran answer is affected by the ill-condition.

[6]

This problem is mildly ill-conditioned, but Octave solution is accurate because, if the answer is substituted into the equation, the left side becomes exactly equal to the right side. On the other hand, if the single precision Fortran answer is substituted, the left side of the first equation becomes 6.9986 while the right side is 7.0000. So the single precision Fortran solution is inaccurate.

[7]

```
%P1_7.m    Hilbert matrix
clear all
n=10
for i=1:n;  for j=1:n;  a(i,j)=1/(i+j-1);  end;  end
b=a*inv(a);
b(:,6:10)
```

Only columns 6 to 10 are printed below (diagonal elements are in bold font) : :Calculation was done on Octtave)
Off-diagonal elements must be zero in correct results.

9.4604e-004 -1.3504e-003 1.4420e-003 -7.0572e-004 1.6785e-004
7.3624e-004 -1.0300e-003 1.1139e-003 -5.9891e-004 1.2875e-004
5.7983e-004 -9.6893e-004 9.5367e-004 -4.8447e-004 1.0967e-004
4.9973e-004 -8.3160e-004 7.6294e-004 -4.0245e-004 9.6798e-005
4.2725e-004 -6.5613e-004 6.7139e-004 -3.4714e-004 8.2016e-005
1.0004e+000 -6.5613e-004 6.8665e-004 -3.6621e-004 8.2493e-005
3.4714e-004 **9.9952e-001** 4.7302e-004 -2.4223e-004 5.8651e-005
4.1199e-004 -5.4169e-004 **1.0006e+000** -3.0327e-004 6.8665e-005
3.3569e-004 -4.8828e-004 5.4932e-004 **9.9976e-001** 6.8665e-005
2.7466e-004 -3.6621e-004 4.8828e-004 -2.4414e-004 **1.0001e+000**

[8]

```
>>eig(a)'
ans =
  -4.502791  -0.040067  5.542858
```

[9]

```
>>a=[1 2 2; -1 -1 4; 3 5 1]; c=poly(a), Roots_are=roots(c)',
Eigenvalues_are=eig(a)'
c =
   1.0000  -1.0000  -25.0000  -1.0000
Roots_are =
   5.542858  -4.502791  -0.040067
Eigenvalues_are =
  -4.502791  -0.040067  5.542858
```

[10]

det(a) = (4)(5) = 20
det(b) = (3)(1)(2) = 6

[11]

(a) 4, 5
(b) 3, 1, 2

141

[12]

       (a) 1/4, 1/5

       (b) 1/3, 1, 1/2

[13]

       (a) Maximum of eignvalues of $a'a$ is 5^2, and that of $(a^{-1})'a^{-1}$ is $(¼)^2$. The 2-norm of a is 5 and that of a^{-1} is ¼. Therefore, the condition number of a equals $(5)*(¼)=1.25$.

       (b) Maximum of eignvalues of $b'b$ is 3^2, and that of $(b^{-1})'b^{-1}$ is 1^2. The 2-norm of b is 3 and that of b^{-1} is 1. Therefore, the condition number of b equals $(3)*(1)=3$.

Problems for Chapter 2

[1]

     From observation of the graph, $y = 2.5x -5$

     Hint: May use **polyfit([0 2],[-5 0],1)** to work on Octave/Matlab

[2]

       (a)   >>c=polyfit([-1 0 2],[-3 1 -1],2)

       c =

       -1.6667 2.3333 1.0000

       (b)   polyfit([-3 0 2],[6 -2 1],2), $y=0.83333x^2 - 0.16667x - 2$

[3]

       Clustered form: $y = (((2x - 4)x + 1)x + 3)$

       Normalized form: $y = x^3 - 2x^2 + 0.5x + 1.5$

       Factorized form:

       >>roots([2 -4 1 3])

       ans =

        1.33127 + 0.70125i

        1.33127 - 0.70125i

        -0.66254 + 0.00000i

       So, $y = 2(x - 1.33127 - 0.70125i)(x - 1.33127 + 0.70125i)(x + 0.66254)$

[4]

       >>polyval([1 -7 4 5], [-5 -2 1 3 5])

       ans =

       -315 -39   3 -19 -25

[5]

       (a) >>roots([2 -4 -6])'

       ans =

        3.00000 -1.00000

       (c) >>roots([1 1 -9 -9])'

       ans =

        3.0000 -3.0000 -1.0000

(d) >>roots([1 0 -2 0 4 -3])'
ans =
 -1.36166-0.79537i   -1.36166+0.79537i
 0.86166-0.68114i   0.86166+0.68114i
 1.00000-0.00000i

[6]

```
%P2_6.m
clear, clf
c=input( 'Input the power coefficients in array like [ 1 2 0 1] with
[ ]: ')
xmin=input('Input lower bound of x in the plot: ')
xmax=input('Input upper bound of x in the plot: ')
n=length(c);
dx=(xmax-xmin)/40;
x = xmin:dx:xmax;
y=polyval(c, x);
plot(x,y)
```

c =
 1 -5 0 1 2

Input lower bound of x in the plot: -3
xmin = -3
Input upper bound of x in the plot: 6
xmax = 6

[7]
 (a) 2, (b) 3, (c) 3

[8]
(a)
c=polyfit([0 1 2],[2 -1 0], 2)
c =
 2.0000 -5.0000 2.0000
(b)
c=polyfit([1 3 5 7],[0 0 0 1], 3)
c =

143

```
    0.020833  -0.187500  0.479167  -0.312500
(c)
c=polyfit([1 3 5 7],[0 1 0 0], 3)
c =
    0.062500  -0.812500  2.937500  -2.187500
```

[9]

```
%P2_9.m
clear, clf
xp=[0 1 2]; yp=[2 -1 0];        % Used for (a): put % if the script is run
                                  for b or c
%xp=[1 3 5 7]; yp=[0 0 0 1];  % to be used for (b)
%xp=[1 3 5 7]; yp=[0 1 0 0];  % to be used for (c)
c=polyfit(xp,yp, length(xp)-1)
x=-10:0.1:10;
y=polyval(c,x);
plot(x,y, 'linewidth',2), hold on
axis([min(xp)-0.5 max(xp)+0.5 -2 2]);
plot(xp,yp, '*','linewidth',2)
plot([0 0],[-10 10]); plot([-10 10],[0 0])
xlabel('X', 'fontsize', 16); ylabel('Y', 'fontsize', 16)
```

(a)   (b)   (c)

[10]

```
%P2_10.m
clear, clf
disp('The following input must be enclosed in single quote signs.')
x=input( 'Input abscissas of data points in array in string: ',"s")
y=input( 'Input ordinates of data points in array in string: ',"s")
x=str2num(x); y=str2num(y);
g = max(x)-min(x);
xmin=min(x)-0.05*g;
xmax=max(x)+0.05*g;
c = polyfit(x,y,length(x)-1)
dx=(xmax-xmin)/40;
xp = xmin-dx:dx:xmax+dx;
yp=polyval(c, xp);
plot(xp,yp, 'linewidth', 2); hold on
plot(x,y,'*', 'linewidth',2)
plot([xmin-10, xmax+10],[0,0])
plot([0, 0],[min(yp)-20, max(yp)+20]); hold off
```

```
dy=0.05*(max(yp)-min(yp));
axis([xmin-dx,xmax+dx,min(yp)-dy,max(yp)+dy])
xlabel('X', 'fontsize', 16)
ylabel('Y', 'fontsize', 16)
```

Test run:
Input abscissas of data points in array: -10 5 0 3 10
x = -10 5 0 3 10
Input ordinates of data points in array: -1 2 -3 2 -3
y = -1 2 -3 2 -3
c =
   0.0025421 -0.0267106 -0.2442125   2.5710623 -3.0000000

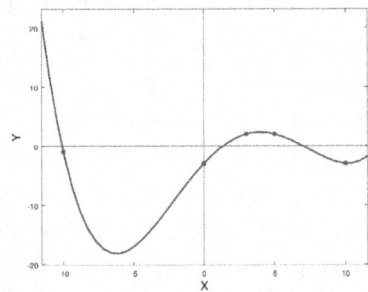

[11]

Data points in array form are:
x=[-5 2 0]
y=[0 0 -2]
c=polyfit(x,y,length(x)-1)

c =
   0.20000   0.60000 -2.00000
or more explicitly
$x = 0.20000x^2 + 0.60000x - 2.00000$

[12]

The polynomial must be in the form,
$y = c(x-a-bi)(x-a+bi) = c((x-a)^2 + b^2) = cx^2 + c2ax + ca^2 + cb^2$
Using the information,
x=1, y=1 : $1 = c + 2ac + ca^2 + cb^2$
x=0, y=5: $y = 5 = ca^2 + cb^2$
Complex conjugate roots are known, that means, the values of a and b
are known. So, the only unknown is c, which can be determined by one
of the equations written above.

[13]

>> roots([2 4 -14 -16 24] ')
ans =
 -3.0000 -2.0000 2.0000 1.0000
Therefore,

——

[14]

y = 2(x+3)(x+2)(x-2)(x-1)

[15]

dy/dx=2ax + b =0, a>0
x = -b/(2a) where y is minimum, for which
y-min = a(-b/2a)^2 – b^2/(2a) + c
    = (b^2/4a)- 2b^2/(4a) + c = (-b^2 + 4ac)/4a
For complex roots only, y-min>0, so -b^2 + 4ac>0 or b^2 - 4ac < 0

[16]

Because a<0, y-max=(-b^2 + 4ac)/4a. For this to be negative, namely (-b^2 + 4ac)/4a<0, we must have -b^2 + 4ac>0 or equivalently b^2 - 4ac < 0, which is the same as the previous problem.

[17]

Regardless to the sign of a, the condition for the two roots to be complex is b^2 - 4ac < 0.

[18]

c = -0.20145 1.43852 -2.74771 5.43700
y = -0.20145x^3 + 1.43852x^2 - 2.74771x + 5.43700

```
%P2_18.m
clear
x = [1.1,  2.3,  3.9,  5.1]';
y = [3.887, 4.276, 4.651, 2.117]' ;
n = length(x);
for k=1:n
yd=y*0; yd(k)=1;
cd(k,:)=polyfit(x,yd,length(x)-1)*y(k);
end
c = sum(cd)


c =
 -0.20145  1.43852 -2.74771  5.43700
```

[19]

```
%P2_19.m
clear,clf
t = [0:pi/4/5:pi/4] ;
c=polyfit(t, sin(t), 5);
x = 0:pi/4/20/4:pi/4;
f = sin(x);
g =  polyval(c, x);
clf, plot(x,f ,'--',x ,g, 'linewidth',4, x, (f - g)*1.0e6, x, 0*x, 'k')
maxerror=max(abs(f-g));
me=  ['|maximum error| = ', num2str(maxerror) ]
text(0.05, 0.7, me ,'fontsize', 16)
text(0.6, max(abs(f-g)*1.0e6)+0.02, 'Error*1.0e6', 'fontsize', 16)
xlabel('X', 'fontsize', 16)
ylabel('Y', 'fontsize', 16)
```

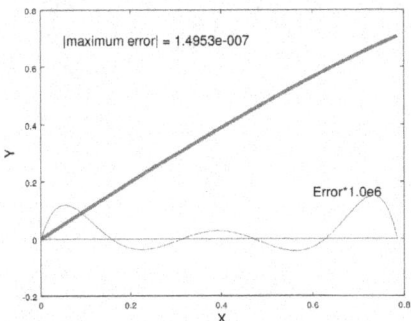

With 5th order polynomial, the maximum error became 1.4953e-7 comparing to 8.2819e-6 of the 4th order polynomial. The trend of increasing errors at the edge subinterval became even greater than the 4th order polynomial fitting.

[20]

```
%P2_20.m
clear, clf
k=7;m=1:k; a = 0; b = pi/4;
t = 0.5*((b-a)*cos((k+0.5-m)*pi/k) + a + b);
c=polyfit(t, sin(t), 6);
x = 0:pi/4/20/4:pi/4;
f = sin(x);
g = polyval(c, x);
clf, plot(x,f,'--',x ,g, 'linewidth',4); hold on
plot( x, (f - g)*1.0e7, x, 0*x, 'k')
maxerror=max(abs(f-g));
me= ['|maximum error| = ', num2str(maxerror) ]
text(0.05, 0.7, me ,'fontsize', 16)
text(0.6, max(abs(f-g)*1.0e7)+0.02, 'Error*1.0e7', 'fontsize', 16)
xlabel('X', 'fontsize', 16)
ylabel('Y', 'fontsize', 16)
```

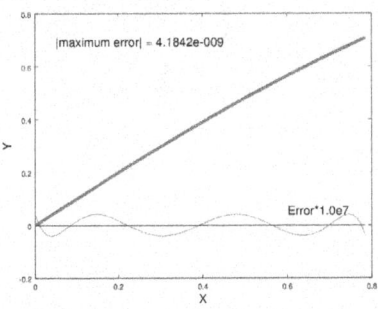

Error distribution is more even than in the results of Problem [19], and the maximum error is significantly lower. This is reasonable because Problem [19] uses 5th order polynomial with evenly distributed points, while 6th order polynomial is used with 7 Chebyshev points here.

Problems for Chapter 3

[1]

```
%P3_1.m
clear
for kase=1:3
if kase==1 f='1./(2+cos(x))'; xlo=0;xhi=pi;fprintf('Case a: \n');end %if
if kase==2, f='log(1+x)./x'; xlo=1;xhi=2;fprintf('Case b: \n');end %if
if kase==3, f='1./(2+(sin(x)).^2)'; xlo=0;xhi=pi/2;fprintf('Case c:
\n');end %if
for n=[2 4 8 16 32 64]
h=(xhi-xlo)/n;
x=xlo:h:xhi;
y=eval(f);
I=h*(sum(y)-0.5*(y(1)+y(length(y))));
%[kase,n,xlo,xhi,I]
fprintf('n=%i, I=%.7f \n', n,I)
end %for
end %for


Case a:
n=2, I=1.8325957
n=4, I=1.8138958
n=8, I=1.8137994
n=16, I=1.8137994
n=32, I=1.8137994
n=64, I=1.8137994

Case b:
n=2, I=0.6160436
n=4, I=0.6147222
n=8, I=0.6143902
n=16, I=0.6143071
n=32, I=0.6142863
n=64, I=0.6142811

Case c:
n=2, I=0.6414085
n=4, I=0.6412749
n=8, I=0.6412749
n=16, I=0.6412749
```

n=32, I=0.6412749
n=64, I=0.6412749

```
%P3_2.m
clear
for kase=1:3
Ib=-999;
if kase==1 f='1./(2+cos(x))'; xlo=0;xhi=pi;fprintf('Case a: \n');end %if
if kase==2, f='log(1+x)./x'; xlo=1;xhi=2;fprintf('Case b: \n');end %if
if kase==3, f='1./(2+(sin(x)).^2)'; xlo=0;xhi=pi/2;fprintf('Case c:
\n');end %if
for n=[2 4 8 16 32 64]
h=(xhi-xlo)/n;
x=xlo:h:xhi;
y=eval(f);
I=h*(sum(y)-0.5*(y(1)+y(length(y))));
%[kase,n,xlo,xhi,I]
if n==2, fprintf('n=%i, I=%.7f \n', n,I);
else fprintf('n=%i, I=%.7f Rmbg=%.7f\n', n,I, (4*I-Ib)/3)
end %if
Ib=I;
end %for
end %for
```

Case a:
n=2, I=1.8325957
n=4, I=1.8138958 Rmbg=1.8076624
n=8, I=1.8137994 Rmbg=1.8137672
n=16, I=1.8137994 Rmbg=1.8137994
n=32, I=1.8137994 Rmbg=1.8137994
n=64, I=1.8137994 Rmbg=1.8137994
Case b:
n=2, I=0.6160436
n=4, I=0.6147222 Rmbg=0.6142818
n=8, I=0.6143902 Rmbg=0.6142795
n=16, I=0.6143071 Rmbg=0.6142793
n=32, I=0.6142863 Rmbg=0.6142793
n=64, I=0.6142811 Rmbg=0.6142793
Case c:
n=2, I=0.6414085
n=4, I=0.6412749 Rmbg=0.6412304
n=8, I=0.6412749 Rmbg=0.6412749
n=16, I=0.6412749 Rmbg=0.6412749
n=32, I=0.6412749 Rmbg=0.6412749
n=64, I=0.6412749 Rmbg=0.6412749

```
%P3_3.m
clear
for kase=1:3
if kase==1 f='1./(2+cos(x))'; xlo=0;xhi=pi;fprintf('Case a: \n');end %if
if kase==2, f='log(1+x)./x'; xlo=1;xhi=2;fprintf('Case b: \n');end %if
if kase==3, f='1./(2+(sin(x)).^2)'; xlo=0;xhi=pi/2;fprintf('Case c:
\n');end %if
for n=[2 4 8 16 32 64]
h=(xhi-xlo)/n;
x=xlo:h:xhi;
y=eval(f);
yend=length(y);
I = h/3*( sum(y)+3*sum(y(2:2:yend-1))+ sum(y(3:2:yend-2)) );
%I=h*(sum(y)-0.5*(y(1)+y(length(y)))));
%[kase,n,xlo,xhi,I]
fprintf('n=%i, Spms:I=%.7f \n', n,I)
end %for
end %for
```

Case a:
n=2, Spms:I=1.7453293
n=4, Spms:I=1.8076624
n=8, Spms:I=1.8137672
n=16, Spms:I=1.8137994
n=32, Spms:I=1.8137994
n=64, Spms:I=1.8137994
Case b:
n=2, Spms:I=0.6143159
n=4, Spms:I=0.6142818
n=8, Spms:I=0.6142795
n=16, Spms:I=0.6142793
n=32, Spms:I=0.6142793
n=64, Spms:I=0.6142793
Case c:
n=2, Spms:I=0.6370452
n=4, Spms:I=0.6412304
n=8, Spms:I=0.6412749
n=16, Spms:I=0.6412749
n=32, Spms:I=0.6412749
n=64, Spms:I=0.6412749

[4]

Simspon's rule result = 18, while the analytical integral becomes
x=2;Ihi=(1/4)*x^4+(1/2)*2*x^2 +5*x;Ilo=0;I=Ihi-Ilo; I=18, and both
agree exactly.
This is because the error term of Simson's rule vanishes when a
polynomial of order 3 is integrated.

———

[5]

Suppose we denote the lowest point of the surface as A, and the highest point B. The projection of A and B on the horizontal x-y plane will be denoted by xA and xB. We assume also the the x-axis starts at xA and passes through xB.   Th z values at A and B are then zA=0 and zB=1m. The y axis starts at point A and perpendicular to the x-axis on the horizontal plane. The surface of the cylinder is a circle and is written as

$$y^\pm = \pm\sqrt{0.5^2 - (x-0.5)^2}$$

Height of the slanted plane on the top is given by
$$z(x) = x$$
So the volume of the cylinder is

$$v = \int_{x=0}^{1} (y^+ - y^-)z(x)dx$$

$$= 2\int_{x=0}^{1} y^+ x dx = 2\int_{x=0}^{1} \sqrt{0.5^2 - (x-0.5)^2}\, x dx$$

```
%P3_5.m
clear; h=0.00001;x=0:h:1;
y = 2*x.*sqrt(0.5^2 - (x-0.5).^2);
I=h*sum(y)-h/2*y(length(y))
```

I = 0.392699068550853

Exact analytical answer = 0.5^3*pi= 0.5^3*pi
    = 0.392699081698724
Agreement is excellent.

[6]

```
%P3_6.m
clear
a=0; b=30;
h=(b-a)/10;
x=a:h:b;
yd=0.1*(30-2*x);
f=sqrt(1+yd.*yd);
L=length(f);
I_trapez= h*(sum(f)-0.5*(f(1)+f(L)))
I_trapezH= 2*h*(  sum(f(1:2:L))-0.5*(f(1)+f(L))  )
I_simps=(4*I_trapez - I_trapezH)/3
```

I_simps = 56.5222204304533 (m)

[7]

```
%P3_7.m
```

```
clear
a=15; b=30;  h=(b-a)/2; v=a:h:b;
f=5400./(8.276*v.^2+2000).*v;  L=length(f);
I_simps=h*(sum(f)+3*f(2))/3
```

I_simps = 291.590064878421 (exact 291.86 m)

[8]

```
%P3_8.m
n=20;
for k=1:4
X=10;h=2*X/n; x = -X:h:X; f=exp(-x.^2)./(1+x.^2);
I=h*sum(f); n=n*2; fprintf(' n=%i  I=%.7f \n', n,I)
end %for
```

```
n=40    I=1.3752304
n=80    I=1.3433530
n=160  I=1.3432934
n=320  I=1.3432934
```

[9]

(a)
```
%P3_9.m
clear
n=20; a=0   ;b=1; func='tan(x)./x.^0.7';
for k=1:4
Z=16;h=2*Z/n; z = -Z:h:Z; x=(a+b+(b-a)*tanh(z))/2;
dxdz=(b-a)./(2*(cosh(z)).^2);
f=eval(func) .* dxdz;
I=h*sum(f); n=n*2; fprintf(' n=%i  I=%.7f \n', n,I)
n=n*2;
end %for
```

I=0. 9063459

(b) Use the same script as for (a) except the last part of 2nd line is
changed to
```
func='exp(x)./sqrt(1-x.*x)';
```
I=3.1043786

[10]

(a)
```
%P3_10.m
clear
hx=0.5; hy=0.5;x=1:hx:2;  y=0:hy:1;
%Trapezoidal
```

```
for i=1:3
Iy(i)=hy*( sin(x(i)+y(1))+ 2*sin(x(i)+y(2))+sin(x(i)+y(3)))/2;
end %for
Itrapez=hx*(Iy(1)+2*Iy(2)+Iy(3))/2
%Simpson's
for i=1:3
Iy(i)=hy*( sin(x(i)+y(1))+ 4*sin(x(i)+y(2))+sin(x(i)+y(3)))/3;
end %for
Isimps=hx*(Iy(1)+4*Iy(2)+Iy(3))/3
```

Itrapez = 0.801390254790869
Isimps = 0.836602270038726
(Both results will agree if the number of data points is increased.)
(b)
Itrapez = 3.96685188921747
Isimps = 4.07456204247797
(Same)

[11]
```
%P3_11.m
clear
hx=2.5; x=2:hx:7;
%Simpson
for i=1:3
     ay=exp(-x(i)); by=log(x(i)); hy=(by-ay)/2;y=ay:hy:by;
     for j=1:3; f(j)=log(1+x(i).*y(j))./x(i); end %for
     Iy(i)=hy*(f(1)+4*f(2)+f(3))/3;
end %for
Isimps=hx*(Iy(1)+4*Iy(2)+Iy(3))/3
```

Isimps = 2.03833101589354

Problems for Chapter 4
[1]
```
%P4_1.m
h=[0.1 0.05 0.01 0.001];
de=cos(1);
for i=1:4
df=(sin(1+h(i))-sin(1))/h(i);
dc=(sin(1+h(i))-sin(1-h(i)))/h(i)/2;
db=(sin(1)-sin(1-h(i)))/h(i);
fprintf(...
' f-dif=%.6f er=%.6f c-dif=%.6f er=%.6f b-dif=%.6f er=%.6f\n' ...
,df,de-df,dc,de-dc, db, de-db); end
```

(f:forward, c:central, b:backward, er=error)
f=0.497364 er=0.042939 c=0.539402 er=0.000900 b=0.581441 er=-0.041138

f=0.519045 er=0.021257 c=0.540077 er=0.000225 b=0.561110 er=-0.020807
f=0.536086 er=0.004216 c=0.540293 er=0.000009 b=0.544501 er=-0.004198
f=0.539881 er=0.000421 c=0.540302 er=0.000000 b=0.540723 er=-0.000421

[2]

```
%P4_2.m
h=[0.1 0.01 0.005 0.001];
de=0.5;
for i=1:4
df=(sqrt(1+h(i))-sqrt(1))/h(i);
dc=(sqrt(1+h(i))-sqrt(1-h(i)))/h(i)/2;
db=(sqrt(1)-sqrt(1-h(i)))/h(i);
fprintf(...
' f=%.6f er=%.6f c=%.6f er=%.6f b=%.6f er=%.6f\n' ...
,df,de-df,dc,de-dc, db, de-db)
end
```

(f:forward, c:central, b:backward, er=error)
f=0.488088 er=0.011912 c=0.500628 er=-0.000628 b=0.513167 er=-0.013167
f=0.498756 er=0.001244 c=0.500006 er=-0.000006 b=0.501256 er=-0.001256
f=0.499377 er=0.000623 c=0.500002 er=-0.000002 b=0.500627 er=-0.000627
f=0.499875 er=0.000125 c=0.500000 er=-0.000000 b=0.500125 er=-0.000125

[3]

```
%P4_3.m
clear
h=[0.1 0.05 0.01 0.001];
hh=0.0001
fddd=(sin(1+2*hh)- 2*sin(1+hh)+2*sin(1-hh)-sin(1-2*hh))/(2*hh^3)
fdd=(sin(1+hh)-2*sin(1)+sin(1-hh))/hh^2
fd=(-sin(1+2*hh)+8*sin(1+hh)-8*sin(1-hh)+sin(1-2*hh))/(12*hh)
for i=1:4
df=(sin(1+h(i))-sin(1))/h(i);
dc=(sin(1+h(i))-sin(1-h(i)))/h(i)/2;
db=(sin(1)-sin(1-h(i)))/h(i);
ef=-0.5*h(i)*fdd;
ec=-(1/5)*h(i)^2*fddd;
eb=0.5*h(i)*fdd;
%fprintf(...
%' f-dif=%.6f er=%.6f c-dif=%.6f er=%.6f b-dif=%.6f er=%.6f\n' ...
%,df,fd-df,dc,fd-dc, db, fd-db)
fprintf(...
' f-et=%.6f er=%.6f c-et=%.6f er=%.6f b-et=%.6f er=%.6f\n' ...
,ef,fd-df,ec,fd-dc, eb, fd-db)
end %for
```

(f-et=forward diff error term, er=exact-f-diff,
c-et=centeral diff error term, er=exact-c-diff
b-et=backward diff error term, er=exact-b-diff)
f-et=0.042074 er=0.042939 c-et=0.001081 er=0.000900 b-et=-0.042074 er=-0.041138
f-et=0.021037 er=0.021257 c-et=0.000270 er=0.000225 b-et=-0.021037 er=-0.020807

f-et=0.004207 er=0.004216 c-et=0.000011 er=0.000009 b-et=-0.004207 er=-0.004198
f-et=0.000421 er=0.000421 c-et=0.000000 er=0.000000 b-et=-0.000421 er=-0.000421
(Observation: error terms are predicting error very well.)

[4]

The error term becomes

$$E = \alpha(0.5hf\,") + (1-\alpha)(hf\,") = (0.5\alpha + 1 - \alpha)hf\,"$$

For $E=0$, we must have $1 - 0.5\alpha = 0$, that is, $\alpha = 2$. Then, the difference approximation becomes

$$f_i' \approx 2(f_i - f_{i-1})/h - (f_i - f_{i-2})/2h = \frac{3f_i - 4f_{i-1} + f_{i-2}}{2h}$$

which is three point backward difference approximation.

[5]

Substituting the error term for the central difference approximation yields

$$E = \alpha(-\frac{1}{12}h^2 f\,''') + (1-\alpha)(-\frac{1}{12}(2h)^2 f\,''')$$

$$= \frac{-\alpha - 4(1-\alpha)}{12} h^2 f\,''''$$

For $E = 0$, $3\alpha - 4 = 0$ must be satisfied, namely $\alpha = \frac{4}{3}$.

Therefore,

$$f_i" \approx \alpha(f_{i+1} - 2f_i + f_{i-1})/h^2 + (1-\alpha)(f_{i+2} - 2f_i + f_{i-2})/(2h)^2$$

$$= \frac{-f_{i+2} + 16f_{i+1} - 30f_i + 16f_{i-1} - f_{i-2}}{12h^2}$$

The above equation is the 5-point central difference approximation for the second derivative.

[6]

The easiest way to derive the difference approximation is to run the script in Appendix D, Algorithm 2: The results are as follows:

Difference Approximation Finder
** Number of points? 3
Input the point indices in row vector form like [x x ... x][-1 0 2]
** Order of difference scheme to be derived ?   1

Difference scheme:
 +(-4.00000/(6.00000 h^1))*f(-1.0h)
 +(3.00000/(6.00000 h^1))*f(0.0h)
 +(1.00000/(6.00000 h^1))*f(2.0h)
Error term

(-2.000/ 6.000)h^2 f''

$$f' \approx \frac{f_2 + 3f_0 - 4f_{-1}}{6h} + E$$

The error term is $E = -\frac{1}{3}h^2 f'''$ with h=0.1.

[7]

$$E = \frac{1}{2}h^2 f'''$$

[8]

See Appendix D, Algorithm 1

[9]

(0.4767-0.2412)/0.5= 0.47100

(0.4547- 2*0.7002+0.9653)/0.5^2 =0.078400

(0.0775-0.1528-0.1573+0.3104)/ 0.5^2=0.31120

Problems for Chapter 5

[1]

```
%P5_1.m        PlotGraph
repeat=1
while repeat
clear, clf;
disp(' Next input must be enclosed by single quote signs ')
f=input( 'Input a function :  ')
xmin=input( 'Input minimum x of the graph:  ')
xmax=input( 'Input maximum x of the graph:  ')
n=input( 'Number of points for plotting:  ')
dx=(xmax-xmin)/n;
xp=xmin:dx:xmax;
dx=(xmax-xmin)/40;
x = xmin:dx:xmax;
y=eval(f);
plot(x,y, 'linewidth', 2); hold on
plot([xmin-10, xmax+10],[0,0])
dy=0.05*(max(y)-min(y));
axis([xmin-dx,xmax+dx,min(y)-dy,max(y)+dy])
xlabel('X', 'fontsize', 16)
ylabel('Y', 'fontsize', 16)
repeat=input('If to repeat the run, input 1, or to stop -1: ')
if repeat==-1, break;end %if
end %while
```

Sample session:
Input a function: ' tan(x)-2*x+0.2'
f = tan(x)-2*x+0.2
Input minimum x of the graph: 0
xmin = 0
Input maximum x of the graph: pi/2-0.2
xmax = 1.3708
Number of points for plotting: 50
n = 50
If to repeat the run, input 1, or to stop -1:

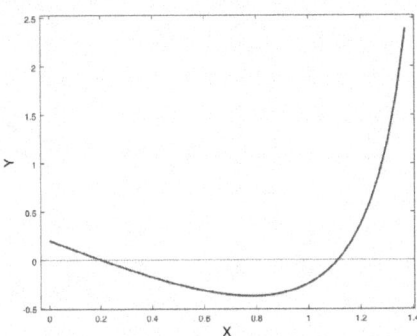

[2]

a   0.2, 1.1
b   0.51, 1.1
c   0.4,   3.2

[3]

(a) Graph 0.8 1.9; Bisection 0.67721   1.90680
(b) Graph 4.0; Bisection 4.01568

[4]

1.2925

[5]

```
%P5_2.m
% Newton iteration  (Script uses definition of function in string)
clear
f='log(x)-0.2*x.*x+1';  %Equation definition
x=0.4        %Initial guess setting
dx=0.001;
for n=1:20
fval=eval(f);
x=x+dx;  fpval=eval(f); x=x-dx;
fderiv=(fpval-fval)/dx;
x=x-fval/fderiv;  diff=-fval/fderiv;
fprintf('n=%f, x=%f, diff=%f\n', n,x, diff)
if abs(diff) <0.0000001, break; end %if
end %for
```

a	Exact	0.202828,	1.111840
b	Exact	0.541948,	1.076464
c	Exact	0.378577,	3.315598

[6]

(a) 0.67721 1.90680

(b) 0.74688

(c) 4.70794

(d) 2.00000

```
clf,clear
%fn='log(1+z)-z.^2';
%fn='tan(z)-2*z+0.2'
%fn='tan(z)-3.5'
%fn='0.5*log(z/3)-sin(z)'
fn='exp(z)-5*z.^2'
%fn='sqrt(z+2)-z'
%fn='log(1+z)-z.^2/10'
x4plot='0:0.1:10'; xinitial=5;
dx=0.001; x=xinitial;
for k=1:20
z=x; f=eval(fn);  plot(x,f,'*'); hold on
xp=x+dx;
fp=log(1+xp)-xp.^2/10;
z=xp; fp=eval(fn);
fd=(fp-f)/dx; x=x-f/fd; if abs(f/fd)<0.000001, break; end %if
end;
residue=f/fd;
fprintf(['Equation="',fn, '=0": Solution=%.5f\n'], x)
fprintf('Total iteration=%i, Residue=%.3e \n', k, residue)
x=eval(x4plot); z=x; f=eval(fn);
plot(x,f)
```

[7]

For graphic plotting:

```
%P5_3.m     Plot of implicit function
clear, clf, cla
xm = -5:0.2:5;  ym = -5:0.2:5; % input for meshgrid
[x, y] = meshgrid(xm, ym);
f = x.*exp(x.*y+0.8)+exp(y.^2)-3;
contour(x,y,f,[0,0])
hold on
f = x.^2 - y.^2 - 0.5*exp(x.*y);
contour(x,y,f,[0,0] )
xlabel('x', 'fontsize',16);
ylabel('y', 'fontsize',16)
```

For solution:

```
%Script for solution
%P5_4f1.m
function f=ff_(x,y)
f = x.*exp(x.*y+0.8)+exp(y.^2)-3;

%P5_4f2.m
function g=gg_(x,y)
g= x.^2 - y.^2 - 0.5*exp(x.*y);
```

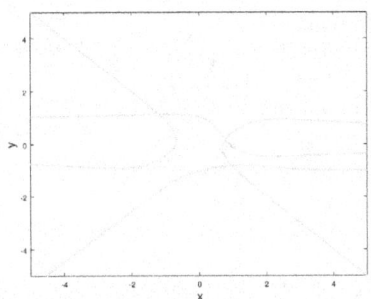

The graphic investigation shows there are three roots, which are approximately (-1.2, 1), (0.9, 0.3), (1, -1).

```
%Continued from the previous page
%P5_4.m      Main part of the script
xest= -1.2;
yest=1;
dx=0.001; dy=0.001
%
for n=1:6
f= ff_(xest,yest);
g=gg_(xest,yest);
fx=( ff_(xest+dx,yest)-f)/dx;
fy=( ff_(xest,yest+dy)-f)/dy;
gx=( gg_(xest+dx,yest)-g)/dx;
gy=( gg_(xest,yest+dy)-g)/dy;
s=-[fx,fy; gx,gy]\[f,g]';
xest=xest+s(1); yest=yest+s(2);
fprintf('n=%f, x=%f, y=%f\n', n,xest,yest)
end
```

With initial guess of (-1.2, 1):
n=1.000000, x=-1.207370, y=1.169300
n=2.000000, x=-1.196674, y=1.142520
n=3.000000, x=-1.196033, y=1.141415
n=4.000000, x=-1.196031, y=1.141412

n=5.000000, x=-1.196031, y=1.141412
n=6.000000, x=-1.196031, y=1.141412

With initial guess of (0.9, 0.3):
n=1.000000, x=0.921808, y=0.580675
n=2.000000, x=0.807273, y=0.326369
n=3.000000, x=0.775313, y=0.198839
n=4.000000, x=0.774857, y=0.172496
n=5.000000, x=0.774976, y=0.171633
n=6.000000, x=0.774976, y=0.171631

With initial guess of (1, -1):
n=1.000000, x=0.987405, y=-0.883492
n=2.000000, x=0.969851, y=-0.849832
n=3.000000, x=0.968791, y=-0.847713
n=4.000000, x=0.968789, y=-0.847709
n=5.000000, x=0.968789, y=-0.847709
n=6.000000, x=0.968789, y=-0.847709

Problems for Chapter 6
[1]

```
%P6_1.m
clear,clf
data=[ ...
1.0      2.0; ...
1.5      3.2; ...
2.0      4.1; ...
2.5      4.9; ...
3.0      5.9; ]
x=data(:,1); y=data(:,2)
c=polyfit(x,y,1);hold on
xp=0:1:4;
yp=polyval(c,xp);
plot(xp,yp,'linewidth',2)
plot(x,y,'*', 'linewidth',2)
xlabel('X','fontsize',16)
ylabel('Y','fontsize',16); hold off
```

[2]

```
%P6_2.m
clear,clf
data=[ ...
0.1      9.9; ...
0.2      9.2; ...
0.3      8.4; ...
0.4      6.6; ...
0.5      5.9; ...
0.6      5.0; ...
0.7      4.1; ...
0.8      3.1; ...
0.9      1.9; ...
1.0      1.1; ...
]
x=data(:,1); y=data(:,2)
%Part 1 Overdetermined equation
a(:,1)=x;a(:,2)=1;
c_overdetermined=(a'*a)\(a'*y)
%Part 2 Using polyfit: Line regression
c=polyfit(x,y,1);hold on
xp=0:1:1;
yp=polyval(c,xp);
plot(xp,yp,'linewidth',2)
plot(x,y,'*', 'linewidth',2)
xlabel('X','fontsize',16)
ylabel('Y','fontsize',16); hold off
c_polyfit=c

c_overdetermined =
 -10.012
  11.027
c_polyfit =
  -10.012   11.027
```

[3]

Taking log of the equation reduces it to a linear equation.

log(y) = log(al) + b*x
logy=logal+be*x

Then, the linear coefficients, logal and be, can be determined as a linear regression problem.

```
%P6_3.m
clear,clf
data=[
0.0129 9.56;
0.0247 8.1845;
0.0530 5.2616;
0.1550 2.7917;
0.3010 2.2611;
0.4710 1.7340;
0.8020 1.2370;
];
logy=log(data(:,2))
x=data(:,1);
c=polyfit(x,logy,1);
logal=c(2);
be=c(1)
al=exp(logal)
plot(x,data(:,2),'*', 'linewidth', 4);hold on
plot(x, al*exp(be*x))
xlabel('X', 'fontsize', 16);
ylabel('Y', 'fontsize', 16);
```

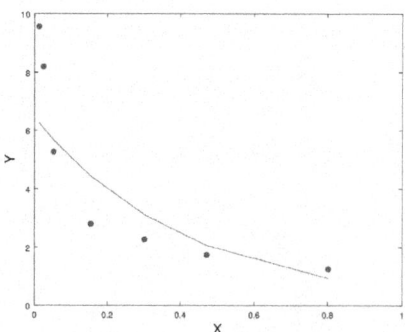

Comment: This fitting is not very successful because of the behavior of the y-values toward x=0. It could be better if a quadratic or cubic polynomial is fitted to x versus log(y).

[4]

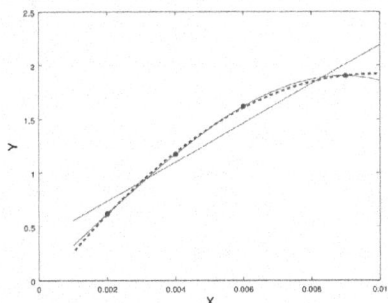

Straight line: linear, curved dot: quadratic, smooth curve: cubic

```
%P6_4.m
clear,clf, hold on
data=[
0.002    0.618;
0.004    1.1756;
0.006    1.6180;
0.009    1.9021
]
xp=0.001:0.0001: 0.010;
x=data(:,1); y=data(:,2);
plot(x,y,'*','linewidth',3)
for k=1:3
c=polyfit(x,y,k)
yp=polyval(c,xp);
if k==1, plot(xp,yp); end %if
if k==2, plot(xp,yp,'--', 'linewidth',2); end %if
if k==3, plot(xp,yp); end %if
end %for
xlabel('X', 'fontsize', 16);
ylabel('Y', 'fontsize', 16);
```

[5]

```
%P6_5.m
clear,clf, hold on
data=[
0.1      0.0000;
0.2      2.1220;
0.3      3.0244;
0.4      3.1399;
0.6      2.8579;
0.7      2.5140;
0.8      2.1639;
0.9      1.8358]
xp=0.:0.01: 1;
x=data(:,1); y=data(:,2);
```

163

```
plot(x,y,'*','linewidth',3)
a=[x*0+1, x, sin(pi*x), sin(2*pi*x)]
c=(a'*a)\(a'*y)
yp=(xp*0+1)*c(1) + xp*c(2) +sin(pi*xp)*c(3)+ sin(2*pi*xp)*c(4);
hold on; plot(xp,yp)
xlabel('X', 'fontsize', 16);
ylabel('Y', 'fontsize', 16);
```

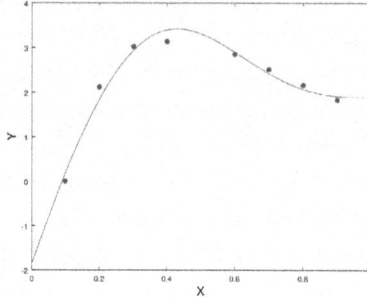

References

[1] S. Nakamura, Numerical Analysis and Graphic Visualization with MATLAB, 2nd Edition, Prentice-Hall, 2002

[2] S. Nakamura, GNU Octave Primer for Beginners, EZ Guide to the Commands and Graphics, Amazon.com, 2015

[3] The MathWorks, MATLAB Student Version Release 12, 2001

[4] J. S. Hansen, GNU Octave, Packet Publishing, 2011

GNU Octave/Matlab Tutorial Series

Volume 1
GNU Octave Primer for Beginners
EZ Guide to the Commands and Graphics
Amazon.com

Volume 2
Foundation of Numerical Analysis
Amazon.com

Volume 3
Numerical Methods for Ordinary Differential Equations
(Forthcoming, Amazon.com)

Volume 4
Numerical Methods for Partial Differential Equations
(Forthcoming, Amazon.com)